SpringerBriefs in Applied Sciences and Technology

More information about this series at http://www.springer.com/series/8884

Iraj Sadegh Amiri · Abdolkarim Afroozeh

Ring Resonator Systems to Perform Optical Communication Enhancement Using Soliton

Springer

Iraj Sadegh Amiri
Department of Physics
Photonics Research Centre
University of Malaya
Kuala Lumpur
Malaysia

Abdolkarim Afroozeh
University Technology Malaysia
Skudai
Malaysia

ISSN 2191-530X   ISSN 2191-5318   (electronic)
ISBN 978-981-287-196-1   ISBN 978-981-287-197-8   (eBook)
DOI 10.1007/978-981-287-197-8

Library of Congress Control Number: 2014947694

Springer Singapore Heidelberg New York Dordrecht London

Printed on acid-free paper

Springer is part of Springer Science+Business Media (www.springer.com)

Contents

Author Biography

Dr. Iraj Sadegh Amiri received his B.Sc. (Hons, Applied Physics) from Public University of Oroumiyeh, Iran in 2001 and is a gold medalist M.Sc. from Universiti Teknologi Malaysia (UTM), in 2009. He was awarded a Ph.D. degree in Nanophotonics in 2014. He has published more than 250 journals/conferences and books in Optical Soliton Communications, Laser Physics, Nanophotonics, Nonlinear fiber optics, Quantum Cryptography, Optical Tweezers, Nanotechnology, Biomedical Physics, and Biotechnology Engineering. Now he is a visiting research fellow at University Technology Malaysia (UTM).

Photonics Research Centre, University of Malaya, 50603 Kuala Lumpur, Malaysia, e-mail: isafiz@yahoo.com

Abdolkarim Afroozeh received his Ph.D. in Photonics from Universiti Teknologi Malaysia (UTM) in 2014. His areas of interests include microring resonators, optical solitons, fiber couplers, and nanowaveguides.

Department of Physics, Fasa University, Fasa, Iran, e-mail: afroozeh155@yahoo.com

Abstract

High capacity and secured optical communication signals are of importance in optical fiber communication. Ring resonators can be used for optical communication. In this book, the security and capacity performed by soliton signals for optical communications are investigated. The Optisystem and MATLAB softwares are used to simulate the results based on the iterative method. Actual data from practical experiments are implemented for generating programming codes. The high capacity of the output signals is obtained by generation of multiple signals, available from channels and large bandwidth. Secured communication is performed using dark soliton pulse and chaotic signal with a spectrum of wavelengths. The chaotic signal, dark soliton, ultra-short pulses, soliton array are investigated for enhancing the security and capacity of optical communication systems. Therefore, the ring resonator system can be used to provide secured and high capacity for optical soliton communication signals using input Gaussian beam, bright and dark soliton pulses.

Chapter 1
Introduction of Soliton Generation

Abstract In communications, a waveguide has a physical structure used to guide the electromagnetic waves. Optical fiber was used widely in the 1960s due to its importance to the communications industry. Ring resonators can be used in many research areas such as optical communications, signal processing, and network security in the nanoscale regime, where they have shown promising applications. Nonlinear ring resonators can be integrated into a single system. The subject of multisoliton light generation is an interesting research technique which can be used to increase the communication channel capacity. High capacity and secured optical communication signals are major concerns in optical fiber communication and signal processing. Nonlinear behavior of the signals inside the ring resonators show interesting phenomena where the secured and high capacity ranges of signals in the form of soliton can be obtained for long distance communication.

Keywords Optical waveguide · Fiber optics · Microring resonators · Security · High capacity soliton signals

Ring resonators can be used in many research areas such as optical communications, signal processing and network security in the nanoscale regime, where they have shown promising applications [1, 2]. Soliton pulse propagating within a Kerr type nonlinear medium is used to generate large bandwidth of signals [3]. By using suitable parameters, for instance, the input power, coupling coefficients, and ring radii, the output signals can be generated and controlled within a series of micro/nano ring resonators [4].

Different configurations of the optical systems can be used to generate secure single and multisoliton pulses coherent. Nonlinear ring resonators can be integrated into a single system [5]. Dark soliton array can be carried out when soliton pulses from single systems mix and combine together, propagating along the fiber [6, 7]. In this book the chaotic signals and dark soliton are investigated in order to confirm the security of the output power. Many approaches can be employed to generate secure signals which we can mention for instance fast and slow light and generation of single photon switching which have good effects in optical communication networks [8, 9].

© The Author(s) 2015

I. Sadegh Amiri and A. Afroozeh, *Ring Resonator Systems to Perform Optical Communication Enhancement Using Soliton*, SpringerBriefs in Applied Sciences and Technology, DOI 10.1007/978-981-287-197-8_1

The capacity of the output signals can be increased when the signal becomes multiple inside different configuration of the ring resonator [10, 11]. The used system can be made of ring resonators, an add-drop filter system or a device known as PANDA. The PANDA ring resonator consists of one add/drop ring resonator which is connected to one or two smaller rings on the right and left side [12–14]. This system is a good design of integrated microring circuit which can be used as interferometer device. Here the secured and multiple signals can be generated and used for transferring of high capacity information along the fiber in long distance communication [15].

The subject of multisoliton light generation becomes very interesting research technique which can be used to increase the communication channel capacity. The secured optical signal is the second advantage of the long distance communication. There are two techniques are used to generate the soliton pulses. First, Pornsuwancharoen et al. [16] have shown that theoretically the soliton pulses can be generated when a pumped soliton pulse within a microring resonator system is used. In this way, the large signal amplification is also achieved. Second, a Gaussian soliton can be generated, in which a simple system configuration can be set up to form the soliton pulse within the medium.

In order to achieve the basic requirement of the multisoliton generation, the intense input light into the system is needed [17, 18]. There are two ways to obtain the requirement. First using high-power light source and second reduce the media length. Here the media represent a ring radius. So far many researchers have worked in both theoretical and experimental works using a common Gaussian pulse for investigation of soliton [19]. Many of the soliton-related concepts in fiber optic are discussed by Agrawal [20]. Generally the use of a Gaussian pulse to form a multisoliton using a fiber ring resonator is recommended [21].

In communications, a waveguide has a physical structure used to guide the electromagnetic waves. It can be used to transmit waves over a wide portion of the electromagnetic spectrum used especially in the microwave and optical frequency ranges. They can be constructed as conductive or dielectric materials based on the working frequency to transfer of power and communication signals.

The theoretical part of the study of waveguide could be analyzed by solving Maxwell's equations, or the electromagnetic wave equation, with boundary conditions based on the properties of the materials and their interfaces. The mode of the propagation within the waveguide depends on the operating wavelength, polarization and the shape and size of the guide. In hollow metallic waveguides, the modes of propagation are defined as a transverse electric TE10 mode for rectangular and TE11 for circular waveguides. The construction of the waveguide has the ability of confining and sustaining the energy of an electromagnetic wave to a specific quite narrow and controllable path [22].

The properties of a closed waveguide are as follows. It guides the electromagnetic wave and has a circular or rectangular cross section [23]. It was electrically conducted walls and can be made hollow or filled with a dielectric material which is good for support a large number of discrete propagating modes in which each discrete mode define the propagation constant for that mode. Here there is no radiation field, where the discontinuities and bends cause mode conversion but not radiation [24].

Optical fiber was used widely in the 1960s due to its importance to the communications industry. The types of dielectric waveguides such as optical fibers were investigated by several people at 1920s, where the most famous of them are Rayleigh, Summerfield and Debye. A dielectric waveguide consists of a dielectric material surrounded by another dielectric material such as air, glass, or plastic having a lower refractive index. Optical fiber is a good example of a dielectric waveguide. The important phenomena such as emission, transmission, detection, amplification, modulation, and switching of light occur inside the fiber optics [25].

The telecommunications revolution occurred at the late 20th Century, where it provides the communications for the Internet. Photonics focuses on communication, where it covers a huge range of science and technology applications, including laser manufacturing, biological and nanotechnology. Ring resonators are interesting devices which can be used for many applications in optical communication and network systems. It has the ability to be linked with other optical components and become a single integrated system. Integrated ring resonators have been used in the last few years and emerged into many applications in optical communication. They are suited for monolithic integration with other components [26].

High capacity and secured optical communication signals are major concerns in optical fiber communication and signal processing. Self Phase Modulation (SPM) causes compression in the propagating pulse. This effect will be balanced by the dispersion of the fiber, where a pulse travels without any change in shape or energy [27, 28]. The highly stable, steady and well localized pulses are the optical solitons. When the soliton pulses are used as an information carrier, the effect of dispersion and nonlinearity balance each other where it can be improved by using the dark solution [29, 30].

There are some limitations of the optical fiber communication such as error detection, signal distortion and crosstalk due to the optical losses and dispersion. One of the solutions to overcome these problems is the use of ultra-short pulses (USP) and transmission soliton pulses under the condition of nonlinear Schrodinger (NLS) type equations. Secured communication can be performed when the chaotic signal which has a spectrum of wavelength is generated. It is considered as a useful signal for private multiple communications in which the chaotic signal and an informational signal are combined by the multiplication or convolution process. The transmission of information can be applied to long distance communication. The transmission link consists of a transmitter and a receiver which are connected to each other using nonlinear fiber optics. In the case of communication security, the dark soliton is a suitable candidate for optical network security application. Due to the low amplitude of the dark soliton pulse, detection of the pulse is very difficult thus it can be used in secured optical communication.

The proposed system to generate the chaotic signals consists of a series of ring resonators or a PANDA ring resonator with variable parameters. The nonlinearity effects split the output signal in which the noisy signals can be generated. The multisoliton pulses are used for high performance network. For long distance communication, the increasing in more soliton channels are required [31, 32]. High communication capacity is performed using more available channels and large bandwidth. In order to generate the multisoliton, a system of micro or nanoring resonator such as a PANDA ring resonator, add/drop system or series of ring resonators are presented.

A waveguide is defined as a structure which guide electromagnetic waves. Depending on the number of propagating modes, different waveguide can be used. Differential geometry of the waveguides can be used to confine energy in one or two dimensions thus slab waveguide and fiber can be used respectively. Usually different waveguides are required to guide different frequencies. For the purpose of optical frequency transmission, the used waveguide is typically optical fiber waveguides [33].

An optical fiber is a thin, flexible, transparent fiber that can be used as a waveguide to transmit light along the length of the fiber. It is suitable for guiding light of high frequency and is used widely in fiber optic communications, which allows transmission of signals over longer distances and at higher bandwidths compared to other forms of communication [34].

Nonlinear behavior of the signals inside the ring resonators show very interesting phenomena where the secured pulses with high capacity ranges can be obtained for long distance communication [35, 36]. In this research study, in general, both experimental and theoretical works can be done to generate the required and applicable results. The theoretical part of this study uses simulation programming to show results by proposing suitable systems consist of ring resonators with appropriate parameters [37, 38]. In order to write the simulation programs, the developed equations for the complicated nonlinear systems are converted to program coding, where they are used to simulate and modeling results via Matlab programming [39, 40]. The basic equations of this research refer to the relation between the electric fields inside single ring in the stationary case extracted from the nonlinear propagation equation. The output power of the ring resonator can be expressed as a function of the number of ring passes in MRR.

To make the proposed systems with practical device, the suitable parameters are used to adjust and provide the output results [41–43]. The radius of the ring has a range of micro scale, where the smaller one can be 800 nm. The coupling coefficient ranges from 0.50 to 0.978 where the effective area of the ring varies from 0.5 to 0.10 μm^2. The nonlinear refractive index of the medium has the range of 10^{-20}–10^{-13} m^2/W, depends on the type of the waveguide [44, 45].

The nonlinear behavior can be used for information security based on the chaotic signals, dark and bright soliton generation and conversion and single photon switching in optical communications [46, 47]. Using this technique, the capacity of the transmission data can be secured and increased when the chaotic packet switching is used. The key advantage of the channel capacity is the multisoliton pulse generation in the multi array or DSA which will be used for high performance communication. The proposed system consists of two or three lines, where each line itself including a series of microring resonators. The systems of micro and nanoring resonators can perform the secured and high capacity signal, where a variety of optical soliton results such as chaotic waves, dark soliton, multisoliton and single photon can be generated. A soliton signal is a good candidate for long distance communication, therefore, increasing of soliton channels is interesting [48, 49]. Large bandwidth of the wavelength from a soliton pulse can be enlarged and stored within a nano-waveguide. The selected light pulse can be trapped and controlled, used for memory generation and large signal amplification [50].

References

1. Amiri IS, Afroozeh A, Nawi IN, Jalil MA, Mohamad A, Ali J, Yupapin PP (2011) Dark soliton array for communication security. Procedia Eng 8:417–422
2. Amiri IS, Ali J (2013) Single and multi optical soliton light trapping and switching using microring resonator. Quantum Matter 2(2):116–121
3. Amiri IS, Nikoukar A, Ali J (2013) Nonlinear chaotic signals generation and transmission within an optical fiber communication link. IOSR J Appl Phys (IOSR-JAP) 3(1):52–57
4. Vázquez C, Vargas S, Pena JMS, Corredera P (2003) Tunable optical filters using compound ring resonators for DWDM. Photonics Technol Lett IEEE 15(8):1085–1087
5. Amiri IS, Alavi SE, Ali J (2013) High capacity soliton transmission for indoor and outdoor communications using integrated ring resonators. Int J Commun Syst. doi:10.1002/dac.2645
6. Amiri IS, Babakhani S, Vahedi G, Ali J, Yupapin PP (2012) Dark-bright solitons conversion system for secured and long distance optical communication. IOSR J Appl Phys (IOSR-JAP) 2(1):43–48
7. Afroozeh A, Amiri IS, Kouhnavard M, Jalil M, Ali J, Yupapin PP (2010) Optical dark and bright soliton generation and amplification. AIP Conf Proc 1341:259–263
8. Malomed BA, Mihalache D, Wise F, Torner L (2005) Spatiotemporal optical solitons. J Opt B: Quantum Semiclassical Opt 7:R53
9. Xu T, Li J, Zhang HQ, Zhang YX, Hu W, Gao YT, Tian B (2008) Integrable aspects and applications of a generalized inhomogeneous N-coupled nonlinear Schrödinger system in plasmas and optical fibers via symbolic computation. Phys Lett A 372(12):1990–2001
10. Amiri IS, Nikoukar A, Shahidinejad A, Ali J, Yupapin PP (2012) Generation of discrete frequency and wavelength for secured computer networks system using integrated ring resonators. In: Computer and communication engineering (ICCCE) conference, Malaysia 2012, IEEE Explore, pp 775–778
11. Amiri IS, Ali J (2013) Nano optical tweezers generation used for heat surgery of a human tissue cancer cells using add/drop interferometer system. Quantum Matter 2(6):489–493
12. Amiri IS, Ali J (2014) Generating highly dark-bright solitons by gaussian beam propagation in a panda ring resonator. J Comput Theoret Nanosci (CTN) 11(4):1–8. doi:10.1166/j ctn.2014.3467
13. Amiri IS, Ali J (2014) Picosecond soliton pulse generation using a PANDA system for solar cells fabrication. J Comput Theoret Nanosci (CTN) 11(3):693–701
14. Amiri IS, Vahedi G, Shojaei A, Nikoukar A, Ali J, Yupapin PP (2012) Secured transportation of quantum codes using integrated PANDA-add/drop and TDMA systems. Int J Eng Res Technol (IJERT) 1(5)
15. Puttacharoen R, Juleang P, Mitatha S, Yupapin PP (2011) Novel optical cryptography using PANDA ring resonator for highly secured communication. Opt Eng 50:075001
16. Pornsuwancharoen N, Dunmeekaew U, Yupapin PP (2009) Multi-soliton generation using a micro ring resonator system for DWDM based soliton communication. Microwave Opt Technol Lett 51(5):1374–1377
17. Amiri IS, Ranjbar M, Nikoukar A, Shahidinejad A, Ali J, Yupapin PP (2012) Multi optical soliton generated by PANDA ring resonator for secure network communication. In: Computer and communication engineering (ICCCE) conference, Malaysia 2012, IEEE Explore, pp 760–764

18. Afroozeh A, Amiri IS, Jalil MA, Kouhnavard M, Ali J, Yupapin PP (2011) Multi soliton generation for enhance optical communication. Appl Mech Mater 83:136–140
19. Deng D, Guo Q (2007) Ince-Gaussian solitons in strongly nonlocal nonlinear media. Opt Lett 32(21):3206–3208
20. Dubreuil N, Baron A, Kroeger F, Trebaol S, Delaye P, Frey R, Agrawal GP (2011) Nonlinear interactions of optical pulses in slow-mode nanowires. In: 2011, IEEE, pp 1–4
21. Harun SW, Shahi S, Sulaimana AH, Ahmada H (2008) Multi-wavelength source based on SOA and EDFA in a ring cavity resonator. Optoelectron Adv Mater, Rapid Commun 2(6):317–319
22. Volodenko AV, Gurin OV, Degtyarev AV, Maslov VA, Svich VA, Senyuta VS, Topkov AN (2011) Transmission characteristics of circular metallic waveguides for terahertz radiation. Quantum Electron 41(9):853–857
23. Singh V, Prasad B, Ojha SP (2004) A comparative study of the modal characteristics and waveguide dispersion of optical waveguides with three different closed loop cross-sectional boundaries. Optik 115(6):281–288
24. Tucker RW, Burton DA, Noble A (2005) Geometry, Sagnac ring lasers and twisted EM modes. Gen Relativ Gravit 37(9):1555–1574
25. Saltiel SM, Sukhorukov AA, Kivshar YS (2005) Multistep parametric processes in nonlinear optics. Prog opt 47:1–73
26. Bogaerts W, De Heyn P, Van Vaerenbergh T, De Vos K, Kumar Selvaraja S, Claes T, Dumon P, Bienstman P, Van Thourhout D, Baets R (2012) Silicon microring resonators. Laser Photonics Rev 6(1):47–73
27. Amiri IS, Ahsan R, Shahidinejad A, Ali J, Yupapin PP (2012) Characterisation of bifurcation and chaos in silicon microring resonator. IET Commun 6(16):2671–2675
28. Amiri IS, Ali J (2014) Characterization of optical bistability in a fiber optic ring resonator. Quantum Matter 3(1):47–51
29. Amiri IS, Raman K, Afroozeh A, Jalil MA, Nawi IN, Ali J, Yupapin PP (2011) Generation of DSA for security application. Procedia Eng 8:360–365
30. Amiri IS, Khanmirzaei MH, Kouhnavard M, Yupapin PP, Ali J (2012) Quantum entanglement using multi dark soliton correlation for multivariable quantum router. In: Moran AM (ed) Quantum entanglement. Nova Science Publisher, New York, pp 111–122
31. Shahidinejad A, Nikoukar A, Amiri IS, Ranjbar M, Shojaei A, Ali J, Yupapin PP (2012) Network system engineering by controlling the chaotic signals using silicon micro ring resonator. In: Computer and communication engineering (ICCCE) conference, Malaysia 2012, IEEE Explore, pp 765–769
32. Afroozeh A, Amiri IS, Bahadoran M, Ali J, Yupapin PP (2012) Simulation of soliton amplification in micro ring resonator for optical communication. Jurnal Teknologi (Sci Eng) 55:271–277
33. Wiederhecker GS, Chen L, Gondarenko A, Lipson M (2009) Controlling photonic structures using optical forces. Nature 462(7273):633–636
34. Bartenev G (1968) The structure and strength of glass fibers. J Non-Cryst Solids 1(1):69–90
35. Alavi SE, Amiri IS, Idrus SM, Supa'at ASM, Ali J (2013) Chaotic signal generation and trapping using an optical transmission link. Life Sci J 10(9s):186–192
36. Alavi SE, Amiri IS, Idrus SM, Ali J (2013) Optical wired/wireless communication using soliton optical tweezers. Life Sci J 10(12s):179–187
37. Jalil MA, Amiri IS, Teeka C, Ali J, Yupapin PP (2011) All-optical Logic XOR/XNOR gate operation using microring and nanoring resonators. Glob J Phys Express 1(1):15–22
38. Teeka C, Songmuang S, Jomtarak R, Yupapin PP, Jalil M, Amiri IS, Ali J (2011) ASK-to-PSK generation based on nonlinear microring resonators coupled to one MZI arm. AIP Conf Proc 1341(1):221–223
39. Amiri IS, Alavi SE, Rahim FJ, Idrus SM, Ali J (2014) Analytical treatment of the ring resonator passive systems and bandwidth characterization using directional coupling coefficients. J Comput Theor Nanosci (CTN)

40. Amiri IS, Rahim FJ, Arif AS, Ghorbani S, Naraei P, Forsyth D, Ali J (2014) Single soliton bandwidth generation and manipulation by microring resonator. Life Sci J 10(12s):904–910
41. Amiri IS, Nikoukar A, Vahedi G, Shojaei A, Ali J, Yupapin PP (2012) Frequency-wavelength trapping by integrated ring resonators for secured network and communication systems. Int J Eng Res Technol (IJERT) 1(5)
42. Amiri IS, Soltanmohammadi S, Shahidinejad A, Ali J (2013) Optical quantum transmitter with finesse of 30 at 800-nm central wavelength using microring resonators. Opt Quantum Electron 45(10):1095–1105
43. Suwanpayak N, Songmuang S, Jalil MA, Amiri IS, Naim I, Ali J, Yupapin PP (2010) Tunable and storage potential wells using microring resonator system for bio-cell trapping and delivery. AIP Conf Proc 1341:289–291
44. Tanaram C, Teeka C, Jomtarak R, Yupapin PP, Jalil MA, Amiri IS, Ali J (2011) ASK-to-PSK generation based on nonlinear microring resonators coupled to one MZI arm. Procedia Eng 8:432–435
45. Amiri IS, Ali J (2013) Controlling nonlinear behavior of a SMRR for network system engineering. Int J Eng Res Technol (IJERT) 2(2)
46. Amiri IS, Ghorbani S, Naraei P, Ali J (2014) Chaotic carrier signal generation and quantum transmission along fiber optics communication using integrated ring resonators. Quantum Matter
47. Amiri IS, Ali J (2013) Data Signal processing via a manchester coding-decoding method using chaotic signals generated by a PANDA ring resonator. Chin Opt Lett 11(4):041901(041904)
48. Amiri IS, Gifany D, Ali J (2013) Long distance communication using localized optical soliton via entangled photon. IOSR J Appl Phys (IOSR-JAP) 3(1):32–39
49. Kouhnavard M, Afroozeh A, Jalil MA, Amiri IS, Ali J, Yupapin PP (2010) Soliton signals and the effect of coupling coefficient in MRR systems. In: Faculty of science postgraduate conference (FSPGC), Universiti Teknologi Malaysia, 5–7 Oct 2010
50. Amiri IS, Nikoukar A, Shahidinejad A, Anwar T, Ali J (2014) Quantum transmission of optical tweezers via fiber optic using half-panda system. Life Sci J 10(12s):391–400

Chapter 2
Mathematics of Soliton Transmission in Optical Fiber

Abstract Ring resonators are suitable for many applications in micro and nano optical communication. Optical soliton is a self-reinforcing solitary wave that maintains its shape while it travels at constant speed. Optical solitons are seen by a cancellation of nonlinear and dispersive effects in the medium which can be a fiber optic. In a Kerr effect medium such as fiber optics, high intensity of light causes a phase delay having similar temporal shape as the intensity. This nonlinear phenomenon occurs for a beam called self-phase modulation (SPM), which is generated by its intensity. Optical Chaos occurs in many nonlinear optical systems. One of the most common examples is a ring resonator. Chaotic behavior has been considered as a nonlinear property in physics, electronics, and communication. When a high-intensity short pulse is coupled to optical fiber, the instantaneous phase of optical pulse rapidly changes through the optical Kerr effect. The SPM and cross-phase modulation (CPM) effects change the phase of the pulse as a function of its intensity. Here, we derive the soliton equations based on solving the nonlinear Schrodinger and Maxwell equations. The main performance characteristics of ring resonators are transmittance, free spectral range, finesse, Q-factor, and group delay, which have been demonstrated.

Keywords Ring Resonators · Solitary Waves · Wavelength Division Multiplexing (WDM) · Self-phase Modulation (SPM) · Cross-Phase Modulation (CPM) · Bifurcation · Chaos · Dispersion · Diffraction · Nonlinear Schrodinger Equation · Maxwell Equation · Add/Drop Ring Resonator · PANDA Ring Resonator

2.1 Ring Resonators

The ring resonators are important devices in nanotechnology which are suitable for many applications in micro and nano optical communication. The nonlinear behaviors of light in a micro and nanoring resonator have been investigated and used for communication security. Moreover, the field buildup inside the ring cavity can be used for all optical signal processing functions based on enhanced nonlinear effects. Here the working principles of optical soliton propagating inside ring resonators are used. The novel concepts of ring resonator devices are given in this chapter.

© The Author(s) 2015
I. Sadegh Amiri and A. Afroozeh, *Ring Resonator Systems to Perform Optical Communication Enhancement Using Soliton*, SpringerBriefs in Applied Sciences and Technology, DOI 10.1007/978-981-287-197-8_2

2.2 Solitary Waves

Optical soliton is a self-reinforcing solitary wave that maintains its shape while it travels at constant speed [1, 2]. Optical solitons are seen by a cancellation of nonlinear and dispersive effects in the medium which can be a fiber optic. Therefore, an optical soliton is a pulse that travels without distortion due to dispersion or other effects. The subject of solitary waves is therefore an important development in the field of optical communications. When solitons pulses are located mutually far apart, each of them is approximately a traveling wave with constant shape and velocity. If two pulses of optical soliton are brought together, they gradually deform and finally merge into a single wave packet. This wave packet, however, can be split into two solitary waves with the same shape and velocity before the collision. Optical solitons are used in telecommunication via Wavelength Division Multiplexing (WDM).

2.3 Wavelength Division Multiplexing (WDM) Technique

Wavelength Division Multiplexing (WDM) is a technique in which simultaneous transmission of signals occurs at different optical wavelength. In some applications of this technique, several optical signals combine, transmit together, and will be separated again based on different arrival times. To handle this technique, optical signals from separating lasers can be combined together or a broadband optical signal from for example a light emitting diode spectrally be sliced into smaller pulses. The use of multiple channels allows increased overall data transmission capacities without increasing the data rates of the single channels, where the time slot per bit must be reduced. Even if the bandwidth of the data modulator is limited, this can be done by using a train of ultra-short pulses (rather than a continuous optical wave) as the input of the modulator [3].

When the transmitted signals reach the receivers, the channel are de-multiplexed by means of optical wavelength filters like Mach-Zehnder interferometers, Febry-Perot resonators or arrayed-waveguide gratings. This technique has the main advantage that the available bandwidth of an optical fiber can be used very efficiently [4]. Therefore DWDM is a technology that employs the properties of refracted light to combine or separate optical signals regarding to their wavelengths within the optical spectrum. DWDM increases the efficiency of using existing fiber by providing multiple optical paths along a fiber. WDM is normally applied to an optical carrier depicted by its wavelength, whereas frequency-division multiplexing (FDM) is applied to a radio carrier. Since wavelength and frequency are connected together through a simple relationship, they describe the same concept.

There are many applications on micro resonators in both fundamental and applied research. One of the existing and emerging micro resonator applications is the wavelength-selective components for WDM systems. The microring and microdisk resonators are flexible building blocks for very large scale integrated photonic circuits, as they are ultra compact (105 devices/cm^2) and can carry out a wide range of optical signal processing functions such as filtering, splitting and combining of light, switching of channels in the space domain, as well as multiplexing and demultiplexing of channels in the wavelength domain. Vernier filter tuning by combining microring resonators of different radii has also been successfully used to suppress non-synchronous resonances of different microrings and extend the resulting filter FSR [5]. Linear arrays of optical micro resonators evanescently coupled to each other can also be used for optical power transfer. This type of coupled-resonator optical waveguide (CROW) has recently been proposed [6] and then demonstrated and studied in a variety of material and geometrical configurations.

2.4 Nonlinear Effect as Self-Phase Modulation (SPM)

In a Kerr effect medium such as fiber optics, high intensity of light causes a phase delay having the similar temporal shape with the intensity. This nonlinear phenomenon occurs for a beam is called self phase modulation (SPM) which is generated by its intensity. This effect refers to nonlinear changes of the refractive index given by

$$\Delta n = n_2 I \qquad (2.1)$$

where, n_2 is the nonlinear index and the optical intensity is shown by I. Therefore this phase shift is a temporal dependence effect, whereas the transverse dependence leads to the effect of self-focusing. In other words, the Kerr effect causes a time-dependent phase shift according to the time-dependent pulse intensity, where a temporally varying instantaneous frequency can be performed shown in Fig. 2.1.

For the initial up-chirped pulse, the SPM leads to spectral broadening, whereas a compression effect occurs for initially down-chirped pulse. In semiconductor lasers high signal intensity will result in reducing the carrier densities, which lead to a modification of the refractive index, where finally the phase will be changed.

Fig. 2.1 Spectral broadening of a pulse due to SPM [7]

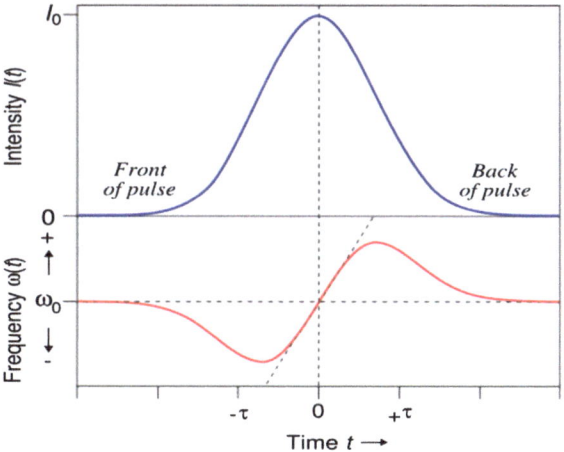

Unlike the way that SPM affects the phase of the propagating pulse such phase changes in semiconductor lasers do not follow the temporal intensity profile. Therefore, this effect is declared for the pulse of picoseconds to a few nanoseconds. SPM is very efficient in mode-locked femtosecond lasers with the Kerr nonlinearity effect medium. In materials with negligible or zero dispersion effects, the nonlinear phase shift is unstable, thus soliton pulse mode is employed which is a result of balancing SPM and dispersion [8, 9]. The intensity of a Gaussian ultrashort pulse at a time (t) can be expressed by

$$I(t) = I_0 \exp(-\frac{t^2}{\tau^2}), \tag{2.2}$$

where I_0 and τ are the peak intensity and pulse duration. In a Kerr type medium, the refractive index is given by

$$n(I) = n_0 + n_2 I, \tag{2.3}$$

where, n_0 and n_2 are the linear and nonlinear refractive indices. Regarding to the intensity profile of the propagating pulse, a time-varying refractive index is given by:

$$\frac{dn(I)}{dt} = n_2 \frac{dI}{dt} = n_2 \cdot I_0 \cdot \frac{-2t}{\tau^2} \cdot \exp(\frac{-t^2}{\tau^2}) \tag{2.4}$$

The phase of the pulse is defined by

$$\phi(t) = \omega_0 t - kx = \omega_0 t - \frac{2\pi}{\lambda_0} \cdot n(I) \cdot L \tag{2.5}$$

where ω_0 and λ_0 are the carrier frequency and wavelength of the pulse, and L is the distance in which the pulse propagates. Due to the change in refractive index, an instantaneous phase shift of the pulse occurs therefore:

$$\omega(t) = \frac{d\phi(t)}{dt} = \omega_0 - \frac{2\pi L}{\lambda_0} \cdot \frac{dn(I)}{dt} \tag{2.6}$$

Therefore

$$\omega(t) = \omega_0 + \frac{4\pi L n_2 I_0}{\lambda_0 \tau^2} \cdot t \cdot \exp(\frac{-t^2}{\tau^2}). \tag{2.7}$$

Thus, the leading edge shifts to lower frequencies where, trailing edge shifts to higher frequencies. In the center portion of the pulse (between $t = \pm \tau/2$), there is an approximately linear frequency shift (chirp) given by:

$$\omega(t) = \omega_0 + \alpha \cdot t \tag{2.8}$$

where α is:

$$\alpha = \frac{d\omega}{dt}\bigg|_0 = \frac{4\pi L n_2 I_0}{\lambda_0 \tau^2} \tag{2.9}$$

In fact in any real medium the dispersion effects are existing and simultaneously will act on the pulse. In normal dispersion the front of the pulse moves faster than the back, broadening the pulse in time, while in anomalous dispersion, the pulse is compressed temporally and becomes shorter. This effect will result generation of ultra-short pulse. If the ultra-short pulse is of sufficient intensity, the spectral broadening process of SPM can balance with the temporal compression due to anomalous dispersion and reach an equilibrium state which is called an optical soliton [10, 11].

2.5 Nonlinear Effect as Cross-Phase Modulation (CPM)

Cross-Phase Modulation (CPM) is a nonlinear optical effect in which one wavelength of light affects the phase of another wavelength of light through the optical Kerr effect. In optical communication networks, CPM can be used as a technique for adding information to a light stream by modifying the phase of a coherent optical beam with another beam through interactions in an appropriate non-linear medium. Thus, it leads to an interaction of laser pulses in a medium in which the measurement of the optical intensity of one pulse can be performed by monitoring a phase change of the other one. This technique is applied in secured optical communication and can also be used for synchronizing two mode-locked lasers using the same gain medium, in which the pulses overlap and experience cross-phase modulation.

2.6 Developments of Bifurcation

The bifurcation is recognized as splitting of a main body of a optical pulse into two parts. A bifurcation occurs when a small change applied to the parameter values (bifurcation parameters) of a system causes a sudden qualitative or a topological change in its behavior. It represents the sudden appearance of a qualitatively different solution for a nonlinear system as some parameter is varied Bifurcations occur in both continuous and discrete systems.

2.7 Developments of Chaos

Optical Chaos occurs in many nonlinear optical systems. One of the most common examples is a ring resonator. Chaotic behavior has been considered as a nonlinear property in physics, electronics and communication [12, 13]. The chaotic communication has recently attracted great attention because of its potential application in secure communication, where it uses a noise-like broadband waveform as a carrier. The chaotic signals can be obtained after propagation of input pulse inside the nonlinear ring resonator [14–16].

Secure data transmission on the basis of chaotic signals is widely investigated in many research works [17, 18]. One of the potential approaches is data transmission using chaotic signals and sequences with modulated parameters. The major concern of this method is a reconstruction of the hidden information from received chaotic signals [19, 20]. Unique properties of chaotic signals serve a basis of application of new radical methods of coding and decoding an information component of chaotic carrying.

2.8   Focussing and Defocussing of Optical Beam

When a laser pulse such as a Gaussian beam propagates in a Kerr type medium along the z-direction, respect to increasing radius away from the center of the beam, the refractive index of the medium will change according to the intensity. Depending on the sign of the Kerr coefficient n_2, with the positive sign, the refractive index decreases with the distance from the beam center and therefore the medium acts like a focusing lens, where the light beam will be compressed towards its center. This phenomena is called self-focusing because the focusing has caused the by light itself [21].

When there are no diffraction effect in the Kerr medium, the light will continue to focus cooperating with increasing in intensity until damage to the material results. If the light is strongly diffracted by the medium, the focusing is countered by the diffractive spreading effects, therefore balancing each other leading to self-trapping of the propagating pulse which is a simple explanation for the existence of the optical spatial soliton. If n_2 is negative, the refractive index increases with the distance from the beam center and the Kerr medium acts like a defocusing lens. This phenomenon is called self-defocusing, where under certain conditions dark spatial soliton pulse can be supported by the medium.

When a light beam in the transverse direction enters an optical Kerr media, the intensity dependent refractive index causes the pulses to be bent through the medium. Therefore, the beam will either experience a defocusing bending effect (if $n_2 < 0$) or a focusing bending effect (if $n_2 > 0$), causes the beam itself will create a self-induced waveguide in the medium.

The most important case is that when $n_2 > 0$. In this case, highly intense beams cause such a strong focusing which can be broken up again, due to strong diffraction effects for very narrow beams, or even due to material damage in the nonlinear crystal. For some situations, however, there exist stationary solutions to the spatial light distribution that exactly balance between the self-focusing and the diffraction of the beam, where it is considered as a balance between two bending effects. The compression can be happened due to self-phase modulation (temporal analogue of self-focusing), where the expansion is due to dispersion. These forces balance and can create a soliton that is localized in the time dimension as well as in the direction of propagation.

2.9 Developments of Dispersion

Dispersion indicates the wavelength dependence of the refractive index in matter, $(dn/d\lambda)$, where n is the refractive index and λ is the wavelength. The spectral region shorter than the zero-dispersion wavelength λ_0 is recognized to a normal dispersion region. A spectral region longer than λ_0 is called an anomalous dispersion region, where the low frequency (or longer wavelength), causes smaller group velocity dispersion (GVD). This phenomenon occurs by interaction between the matter and light because an electromagnetic signal propagating within a physical medium is degraded due to having different propagation velocities of its various wave components such as frequency. Therefore material dispersion causes different wavelengths to travel at different speed because the refractive index of the fiber core varies with wavelength.

The part of light, which travels in the cladding of the fiber having different refractive indices, propagate through it at different speed to the core due to the waveguide dispersion. The Material and waveguide effects are combined to give an overall effect called "chromatic dispersion", which is proportional to the square of the transmitted bit rate in an optical network. Chromatic dispersion is to emphasize its wavelength dependence nature, while group velocity dispersion (GVD) to emphasize the role of the group velocity. Any information-carrying signal included components from a range of wavelength. The group velocity of a signal is function of wavelength, therefore each spectral component travel independently and experiences a group delay, which ultimately results in pulse broadening [22]. Thus, overlapping of pulses will happen and after a certain amount of overlap, adjacent pulses cannot be identified at the receiver and error occurs.

Waveguide dispersion is only important in single mode fibers. It is caused by the fact that some light travels in the fiber cladding compared to most light travels in the fiber core. Since fiber cladding has lower refractive index than fiber core, light ray that travels in the cladding travels faster than that in the core. Waveguide dispersion is also a type of chromatic dispersion. While the difference in refractive indices of the single mode fiber core and cladding is minuscule, they can still become a factor over greater distances. It can also combine with material dispersion to create a nightmare in single mode chromatic dispersion.

2.10 Developments of Diffraction

In a single-mode fiber, the propagating light is confined to a circular area with a diameter of the order of micron size corresponds to the diameter of the fiber core. In the case of free space, this strong confinement would lead to a large divergence angle. In a fiber optic the spatially varying refractive index has a focusing effect which exactly balances diffraction effect for certain fiber modes. In a single-mode fiber, the waveguide effect is just strong enough to balance diffraction.

Thus, the diffraction becomes stronger for smaller modes, where the fibers with small mode areas require a larger numerical aperture (NA) for thoroughly guiding. The large mode area fibers require a small NA if the guidance of higher-order modes is to be kept. Therefore, single-mode fibers with large mode area are based on the balance of two weak effects which are diffraction and correspondingly guiding effect from the waveguide structure. Therefore, any additional effects, such as those from bending the fiber can have a strong impact in the form of bend losses.

2.11 Transmission Theory

If we introduce a pure monochromatic wave at frequency ω_0 into a length of optical fiber, the magnitude of the electric field vector associated with the wave would be given by

$$|E(r,t)| = J(x,y)\cos(\omega_0 t - \beta(\omega_0)z), \tag{2.10}$$

here the z is propagation direction and $J(x, y)$ is the distribution of the electric field along the fiber cross section and is determined by solving the wave equation. For the fundamental mode, the longitudinal component is of the form

$$E_z = 2\pi J_l(x,y)\exp(i\beta z), \tag{2.11}$$

Due to the cylindrical symmetry of the fiber, $J_l(x,y)$ is a function only of $\rho = \sqrt{x^2 + y^2}$ which is expressed in terms of Bessel functions. The transverse component of the fundamental mode is of the form

$$E_x(E_y) = 2\pi J_t(x,y)\exp(i\beta z), \tag{2.12}$$

where again $J_t(x,y)$ depends only on $\sqrt{x^2 + y^2}$. Thus, for each of the solutions corresponding to the fundamental mode, we can write

$$E(r,\omega) = 2\pi J(x,y)\,e^{i\beta(\omega)z}e(x,y), \tag{2.13}$$

where, $J(x,y) = \sqrt{J_l(x,y)^2 + J_t(x,y)^2}$ and the e is the unit vector along the direction of $E(r,\omega)$. In this equation, we have written β as a function of ω to emphasize this dependence. In general, the dependence of the J and e on ω can be neglected for pulses whose spectral width is much smaller than their center frequency. This condition is satisfied by pulses used in optical communication systems.

Consider a pulse whose shape, or envelope, is described by $A(z, t)$ and whose center frequency is ω_0. Assume that the pulses have narrow spectral width. Therefore, most of the energy of the pulse is concentrated in a frequency band whose width is negligible compared to the center frequency ω_0 of the pulse. This assumption is usually satisfied for most pulses used in optical communication systems. Thus, the electric field vector associated with such a pulse is

$$|E(r,t)| = J(x,y)\,\Re\left[A(z,t)e^{-i(\omega_0 t - \beta_0 z)}\right], \tag{2.14}$$

where $\Re$ denotes the real part. Here β_0 is the value of the propagation constant β at the frequency ω_0 and $J(x, y)$ has the same significance as before. Considering the pulse envelope $A(z, t)$ to be complex valued and regarding to the phase shifts it can be written,

$$A(z, t) = |A(z, t)| \exp(i\varphi A(z, t)), \tag{2.15}$$

the phase of the pulse is given by

$$\varphi(t) = \omega_0 t - \beta_0 z - \varphi_A(z, t) \tag{2.16}$$

Here we have also assumed that the pulse is obtained by modulating a nearly monochromatic source at frequency ω_0. We can derive the following partial differential equation for the evolution of the pulse shape $A(z, t)$ [22] as:

$$\frac{\partial A}{\partial z} + \beta_1 \frac{\partial A}{\partial t} + \frac{i}{2} \beta_2 \frac{\partial^2 A}{\partial t^2} = 0, \tag{2.17}$$

where, $\beta_2 = \left. \frac{\partial^2 \beta}{\partial \omega^2} \right|_{\omega=\omega_0}$. If β is a linear function of ω, $\beta_2 = 0$, then arbitrary pulse shapes propagate without change in shape (and at velocity $1/\beta_1$). In other words, if the group velocity is independent of ω, n_0 broadening of the pulse occurs. Thus β_2 is the key parameter governing group velocity or chromatic dispersion. It is termed the group velocity dispersion parameter or, simply, GVD parameter.

2.12 Effects of Nonlinear Condition in Optical Fiber

When a high-intensity short pulse is coupled to optical fiber, the instantaneous phase of optical pulse rapidly changes through the optical Kerr effect. Self-phase modulation (SPM) and cross-phase modulation (CPM) effects change the phase of the pulse as a function of its intensity. For a monochromatic wave the refractive index becomes intensity dependent in the presence of SPM, where for non-monochromatic pulses with envelope A propagating in optical fiber, this relation must be modified so that the frequency and intensity-dependent refractive index is now given by

$$\widehat{n}(\omega, E) = n(\omega) + \bar{n}|A|^2/A_{eff} \tag{2.18}$$

here, $n(\omega)$ is the linear refractive index, which is frequency dependent because of the chromatic dispersion, but also intensity independent, $\bar{n}$ is the nonlinear index coefficient and A_{eff} is the effective cross-sectional area of the fiber, typically $50\ \mu m^2$. The expression for the frequency and intensity-dependent propagation constant is now given by

$$\widehat{\beta}(\omega, E) = \beta(\omega) + \omega \cdot \bar{n}|A|^2/c \cdot A_{eff} \tag{2.19}$$

The unit for the nonlinear index coefficient is $\mu m^2/W$, therefore the intensity of the pulse $|A|^2$ must be expressed in watts (W). For more convenience, we assume that $\gamma = \omega \cdot \bar{n}/c \cdot A_e = 2\pi \cdot \bar{n}/\lambda \cdot A_e$, therefore, $\widehat{\beta} = \beta + \gamma|A|^2$. Thus γ stands the same relationship to the propagation constant β as the nonlinear index coefficient $\bar{n}$ does to the refractive index n. Hence, we call γ the nonlinear propagation. Considering the intensity dependence of the propagation constant, Eq. 2.91 must be modified to

$$\frac{\partial A}{\partial z} + \beta_1 \frac{\partial A}{\partial t} + \frac{i}{2}\beta_2 \frac{\partial^2 A}{\partial t^2} = i\gamma \cdot |A|^2 \cdot A \qquad (2.20)$$

The term of $\frac{i}{2}\beta_2 \frac{\partial^2 A}{\partial t^2}$ incorporating the effect of chromatic dispersion, where, the term $i\gamma \cdot |A|^2 \cdot A$ incorporates the intensity-dependent phase shift. Here, the combined effects of chromatic dispersion and SPM on pulse propagation can be analyzed using this equation as the starting point. Regarding to new variable parameter such as $\tau = (t - \beta_1 z)/T_0, \xi = z/L_D = z|\beta_2|/T_0^2, U = A/\sqrt{P_0}$, Eq. (2.90) can be written as

$$i\frac{\partial U}{\partial \xi} - \frac{\text{sgn}(\beta_2)}{2}\frac{\partial^2 U}{\partial \tau^2} + N^2 |U|^2 U = 0, \qquad (2.21)$$

where

$$N^2 = \gamma P_0 L_D = \frac{\gamma P_0}{|\beta_2|/T_0^2}. \qquad (2.22)$$

Equation (2.25) is called the nonlinear Schrödinger equation (NLSE). Here, the pulse propagates with velocity β_1 (in the absence of chromatic dispersion) and $(t - \beta_1 z)$ is the time axis in a reference frame moving with the pulse. The variable τ is the time in this reference frame but in units of T_0, which is a measure of the pulse width. The variable ξ measures distance in units of the chromatic dispersion length which is defined as $L_D = T_0^2/|\beta_2|$. The quantity P_0 represents the peak power of the pulse, and thus U is the envelope of the pulse normalized to have unit peak power. The quantity of $1/\gamma P_0$ also has the dimensions of length what we call it the nonlinear length and can be shown by L_{NL}.

The nonlinear length serves as a convenient normalizing measure for the distance z in discussing nonlinear effects, just as the chromatic dispersion length does for the effects of chromatic dispersion. Thus for pulses propagating over distances $z \ll L_{NL}$, the effect of SPM on pulses can be neglected. Then we can write the quantity N introduced in the NLSE as $N^2 = L_D/L_{NL}$. If the compression of an optical pulse due to the SPM balances with and cancels the pulse broadening caused by the dispersion, the optical pulse propagates through the fiber while maintaining its original pulse shape. This is called an optical soliton.

2.13 Soliton Solving Based on Nonlinear Schrödinger Equation

The starting point for the analysis of temporal soliton is the time-dependent wave equation for the spatial envelopes of the electromagnetic fields in optical Kerr-media, here for simplicity taken for linearly polarized light in isotropic media,

$$(i\frac{\partial}{\partial z} + i\frac{1}{v_g}\frac{\partial}{\partial t} - \frac{\beta}{2}\frac{\partial^2}{\partial t^2})\vec{A}_\omega(z,t) = -\frac{\omega n_2}{c}\left|\vec{A}_\omega(z,t)\right|^2\vec{A}_w(z,t), \qquad (2.23)$$

where, as previously, $v_g = (\frac{dk}{d\omega})^{-1}$, is the linear group velocity, and the $\beta = \frac{d^2 k}{d\omega^2}$ introduced for the second order linear dispersion of the medium. For the intensity-dependent refractive index $n = n_0 + n_2|E_\omega|^2$. Since we here are considering wave propagation in isotropic media, with linearly polarized light (for which no polarization state cross-talk occur), this equation is conveniently taken in a scalar form as

$$(i\frac{\partial}{\partial z} + i\frac{1}{v_g}\frac{\partial}{\partial t} - \frac{\beta}{2}\frac{\partial^2}{\partial t^2})A_\omega(z,t) = -\frac{\omega n_2}{c}|A_\omega(z,t)|^2 A_\omega(z,t), \qquad (2.24)$$

where it consists of three terms that interact. The first two terms contain first order derivatives of the envelope, and these terms can be seen as the homogeneous part of a wave equation for the envelope, giving travelling wave solutions that depend on the other two terms, which rather act like source terms. The third term contains a second order derivative of the envelope, and this term is also linearly dependent on the dispersion β of the medium, that is to say, the change of the group velocity of the medium with respect to the angular frequency ω of the light. This term is generally responsible for smearing out a short pulse as it traverses a dispersive medium. Finally, the fourth term is a nonlinear source term, which depending on the sign of n_2 will concentrate higher frequency components either at the leading or trailing edge of the pulse [23, 24].

2.14 Ordinary Solitons Generation

The Nonlinear Schrödinger equation, or NLS can be written in another form of

$$-i\frac{\partial u}{\partial z} = \frac{1}{2}\frac{\partial^2 u}{\partial t^2} + |u|^2 u, \qquad (2.25)$$

which is Maxwell's equations, adapted to field propagation in single-mode optical fiber. In a single-mode fiber, there is only one possible spatial behavior in the transverse dimensions x and y, so that we need to deal only with appropriate averages of the field quantities over those dimensions. Thus, the NLS equation involves only distance along the propagation direction, z, and time, t. The absolute magnitude of u represents the amplitude envelope of the pulse and the complex quantity $u(z,t)$ is proportional to the light field and the mean time of flight to location z so that the pulse always remains in view.

The first term on the right of the NLS equation, the one involving the second derivative with respect to time, describes the effects of chromatic dispersion. It is important to note that this linear term, when acting by itself, does nothing to change the frequency spectrum of the pulse. It serves only to broaden (or narrow) the pulse in time. The second term on the right of this equation is the nonlinear term. Note that it is just the pulse's intensity envelope times u itself. It is based on the fact that the index of refraction, is dependent on the light intensity. It is important to note that this term, when acting by itself, does nothing to change the pulse shape in time. It serves only to broaden (or narrow) the pulse in the frequency domain. The major problem is the power loss in the fiber. It occurs primarily due to the absorption and scattering. When loss or gain is taken into account, the NLS equation becomes

$$-i\frac{\partial u}{\partial z} = \frac{1}{2}\frac{\partial^2 u}{\partial t^2} + |u|^2 u - i\frac{1}{2}\alpha u, \tag{2.26}$$

where a positive or negative value of α implies, respectively, gain or loss. The factor of one-half in the α term causes α itself to represent power (or energy) loss or gain per unit length. This equation is one of the few that support solitons. Solitons are a class of solitary wave pulses that can pass through one another with no scattering. This equation has a special solution of

$$u(z,t) = \sec h(t)\exp(\frac{iz}{2}), \tag{2.27}$$

which is known as the fundamental soliton. It has a shape similar to the more familiar Gaussian function, but it has a narrower peak and broader wings. Since the phase term is not dependent on t, the soliton is a completely nondispersive pulse; therefore, its shape does not change with z. This invariance with propagation occurs, for the soliton, since the dispersive and nonlinear terms of the NLS equation cancel each other's effects, causing only a phase shift of the whole pulse. Therefore, the dispersive term, which affects the pulse only in the time domain, can cancel the nonlinear term, which affects the pulse only in the frequency domain. Over very short distances both terms modify only the phase of the pulse [25, 26]. Regarding to the solution of the NLS equation in the regime of $\beta < 0$, a temporal soliton solution can be obtained when the pulse $u(z, t)$ has the initial shape of

$$u(0,t) = N\sec h(t), \tag{2.28}$$

where $N \geq 1$ is an integer number. Depending on the value of N, solitons of different order can be formed, where the fundamental soliton is given for $N = 1$. For higher values of N, the solitons are called "higher order solitons". When $N = 1$, therefore

$$u(z,t) = \sec h(t)\exp(iz/2) \tag{2.29}$$

This solution is called "bright solitons". Bright soliton is characterized as a localized intense peak above a continuous wave (CW) background. This type of optical soliton can be when the group velocity dispersion is negative ($\frac{d^2 k}{d\omega^2}$) or in other words, in the presence of anomalous dispersion. Since the CW plane-wave solution is unstable in self-focusing Kerr medium, bright solitons are a good candidate.

There are a lot of techniques are used to secure secret data or information. Communication security has become the popular technique in modern communication requirement, where the idea of using a dark soliton to be a communication carrier where the recovery can be retrieved by the dark-bright soliton conversion [27–29].

Dark soliton can be used to perform the communication transmission for security, whereas the required information can be retrieved by the dark-bright soliton conversion. Dark soliton is distinguished by being made from a localized reduction of intensity compared to a more intense continuous wave background. The linear polarized dark soliton can be formed in all normal dispersion fiber laser mode-locked by the nonlinear polarization rotation method and can be rather stable. Dark soliton is one of the soliton properties where the soliton amplitude vanished during the propagation in transmission line, thus the dark soliton detection is extremely hard. To date, several papers have investigated the dark soliton behavior [30, 31] and one interesting result is that the dark soliton can be converted to into bright soliton and finally detected. This means that the dark soliton penalty can be used as a communication carrier [32]. Another solution which gives another type of soliton called dark solitons, is expressed as

$$u(z, t) = [\eta \tan h(\eta(t - \kappa z)) - i\kappa] \cdot \exp(iu_0^2 z), \tag{2.30}$$

where u_0 is the normalized amplitude of the continuous-wave background, ϕ is an internal phase angle in the range of $0 \leq \varphi \leq \pi/2$, $\eta = u_0 \cos \phi$ and $\kappa = u_0 \sin \phi$. A dark soliton can be realized as a black soliton (for $\varphi = 0$), which drops down to zero intensity in the middle of the pulse, and the grey soliton (for $\varphi = 0$), which do not drop down to zero. The black soliton, the solution for $\varphi = 0$ takes this simpler form of

$$u(z, t) = u_0 \tan h(u_0 t) \exp(iu_0^2 z) \tag{2.31}$$

Another important difference between the bright and the dark soliton is that the dark soliton propagates through the internal phase angle u_{2z} with a velocity which depends on the amplitude, where the bright soliton propagates with the same velocity.

2.15 Soliton Solving Based on Maxwell Equation

Considering the Maxwell equations, following assumptions can be applied to these equations, leading to We start from the Maxwell equations with the following assumptions:

- Propagation of electric field in the propagation direction can be negligible compared to the transverse electric field.
- The monochromatic electro-magnetic wave is applied, where the frequency is ω.
- The transverse electric field is given by $E(x, y, z, t) = \psi(x, y, z) \exp[i(\omega t - kz)]$, where $\psi(x, y, z)$ changes much slower than $\exp[i(\omega t - kz)]$.
- The refractive index will be changed, where the change is smaller than one.

In the Kerr-type materials, the intensity dependent refractive index is followed by $n(I) = n_0 + n_2 I$. Considering the above assumptions on the Maxwell equation, the nonlinear Schrödinger equation can be derived as

$$\frac{\partial}{\partial z}\psi(x,y,z) = [\frac{i}{2k}(\frac{\partial^2}{\partial x^2} + \frac{\partial^2}{\partial y^2}) + \frac{ik \times n_2}{n_0}|\psi(x,y,z)|^2]\psi(x,y,z), \quad (2.32)$$

where, z is the propagation direction, k is the wave number, and ψ is the amplitude of the electric field. Particular interest which is soliton equation can be derived and expressed by

$$\psi = \sqrt{\psi_0} \sec h[\frac{x}{W_0}] \exp[i(\frac{z}{4z_0})], \quad (2.33)$$

where, $W_0 = \sqrt{\frac{2n_0}{n_2}} \times \frac{1}{k\psi_0}$ is the beam width and $z_0 = \frac{kW_0^2}{2}$ is the diffraction length. The intensity of this equation can be defined by ($I(x,y) = |\psi|^2$) which is independent of z direction. The beam width (W_0) is a function of the peak intensity, where it is wider for lower peak intensities. Therefore, these solutions lead to come out with wave equation which will maintain their shape and size invariant along the propagation direction. If the solution is represented only a state of stationary propagation then it can be considered as a soliton [33–35]. Most phenomenon or nonlinear effects in an optical fiber occur when short pulses with widths ranging from a few ns to a few fs are input into the system, where both dispersive and nonlinear effects contribute to determine the shape and spectrum of the propagating pulses. In 1995, Agrawal could solve the electric field wave equation and derive the governing equation for a light pulse propagating in an optical fiber. This equation is given by

$$\frac{\partial\psi}{\partial z} = (\frac{-i\beta_2}{2}\frac{\partial^2}{\partial T^2} + i\gamma|\psi|^2)\psi \quad (2.34)$$

where ψ and β_2 are the amplitude of the electric field and second-order dispersion of the wave number. T and γ are the temporal profile with the same velocity with pulse group velocity ($T = t - z/v_g$) and nonlinearity coefficient of the medium respectively. Solution of this equation gives solitary waves that keep their profile unchanged when propagating. Considering a negative sign of β_2 and the pulse width of T_0, the solitary solution can be obtained expressed by

$$\psi = \sqrt{\psi_0} \sec h[\frac{T}{T_0}] \exp[i(\frac{z}{2L_D})] \quad (2.35)$$

where ψ_0 is the peak power, $L_D = T_0^2/|\beta_2|$ is the dispersion length of the pulse. This solution, which describes a soliton pulse, shows a pulse that keeps its temporal width invariant as it propagates and hence is called a temporal soliton. It should be taken into account that for any given pulse width, T_0, the peak intensity can be expressed by ($\frac{|\beta_2|}{\gamma T_0^2}$) and it will make the pulse propagating as a solitary wave. The temporal power profile for this soliton is given by

$$I(t,z) = |\psi|^2 = \psi_0 \sec h^2[\frac{t - z/v_g}{T_0}]. \quad (2.36)$$

For the fiber soliton, a balance should be achieved between the dispersion length $L_D = T_0^2/|\beta_2|$ and the nonlinear length $L_{NL} = 1/\gamma\psi_0$, which are the length scales over which disperse or nonlinear effects make the beam become wider or narrower. For a soliton, there is balance between the two and hence $L_D = L_{NL}$, which leads to:

$$T_0^2 \psi_0 = \frac{|\beta_2|}{\gamma} = const, \qquad (2.37)$$

which entails that variations in the power can be compensated by changes in the pulse width and vice versa. This property is responsible for the strength of this type of solitons. Optical excitation having similar parameters to the soliton parameters usually develops with propagation into a stable soliton. For the spatial soliton case for simplicity we will reduce the NLS equation based on the Maxwell assumption to only two dimensions (x, z) which gives

$$\frac{\partial \psi}{\partial z} = [\frac{-i}{2k}(\frac{\partial^2}{\partial x^2} - i\frac{k \times \Delta n_0}{n_0})|\psi|^2]\psi \qquad (2.38)$$

Solution of this equation gives the

$$\psi = \sqrt{\psi_0} \sec h[\frac{x}{W_0}] \exp[i(\frac{z}{4z_0})], \qquad (2.39)$$

Therefore, this solution describes a beam profile which is uniform in y, bell-shaped in x, and propagating in z with stable shape. The intensity structure in x is equal to the mode of a graded-index waveguide with refractive index expressed by

$$n(x) = n_0(1 + \frac{\sec h^2(x/W_0)}{k^2 W_0^2}), \qquad (2.40)$$

Thus, while it propagates, this beam will disintegrate in $\hat{y}$, and some spatial frequency will start to form in $\hat{y}$ due to modulation. This instability can be eliminated by applying the input beam into a slab waveguide made of Kerr nonlinear material.

In nonlinear optics subject, soliton can be classified to temporal or spatial, depending on whether the confinement of light occurs in time or space. Temporal soliton is introduced to optical pulse which maintain its shape, whereas spatial soliton represent self-guided beams that remain confined in the transverse directions orthogonal to the direction of propagation. A spatial soliton is formed when self-focusing of an optical beam balances its natural diffraction-induced spreading. In contrast, it is the self-phase modulation that counteracts the natural dispersion-induced broadening of an optical pulse and leads to the formation of a temporal soliton. Therefore, whether the soliton pulse is of temporal or spatial type, they evolve from a nonlinear change in the refractive index of an optical material induced by the light intensity. This phenomenon is known as optical Kerr effect and optical material that exhibit this property is known as a Kerr medium.

On the other hand, the effect of anomalous dispersion (with $\beta < 0$) and the effect of a nonlinear, intensity dependent refractive index (with $n_2 > 0$) are opposite of each other. The transmission bit rate in optical fibers is limited strongly by group velocity dispersion. It is because generated impulses have a non-zero bandwidth and the medium they are propagating through has a refractive index that depends on frequency (or wavelength). When these two effects (dispersion and nonlinear effects) combined, for example considering pulse propagation in a medium, giving a pulse that can propagate without altering its shape. This is the basic principle of the temporal soliton.

2.16 System of Micro Ring Resonator

A fiber optic ring resonator consists of a waveguide in a closed loop which is coupled to one or more input/output (or bus) waveguides [36–38]. A simple microring resonator is shown in Fig. 2.2.

Ring resonator provides traveling wave procedure, unlike the standing wave characteristic of Fabry-Perot resonators (F-P). Ring resonator can be considered as an interferometer device, which resonates for light whose phase change is an integer multiple of 2π after each trip around the ring [39–41]. Part of light that does not contribute this resonant condition will be transmitted through the bus waveguide. Signal loss occurs when light is transmitted through the fiber, especially over long distances such as undersea cables. The expression for the resonant wavelengths of the ring is very similar to that of the F-P and is given by

$$\lambda_r = \frac{2\pi R n_{eff}}{m} \tag{2.41}$$

where, R is the ring radius constructed with circular waveguide and m is an integer. In this situation the device will act as a phase filter where all wavelengths are transmitted and the resonant wavelengths, having also traversed the ring, acquires a phase change. To capture or separate the resonant wavelengths from the rest, an additional waveguide as an output bus, can be positioned on the opposite side of the ring. In this case the ring resonator is known as an add/drop filter system. The key performance parameters of the ring resonator include the *FSR*, the *ER*, and the finesse. The expression for the *FSR* of a ring resonator is given by

Fig. 2.2 Schematic diagram for a ring resonator coupled to a single waveguide

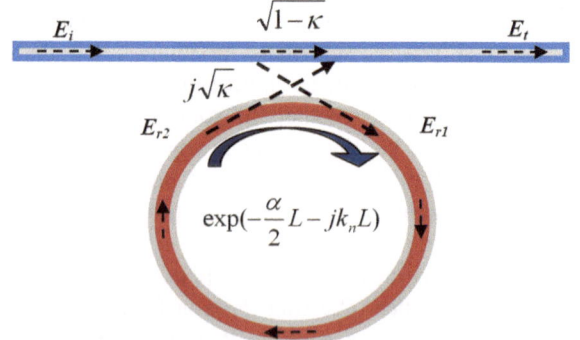

$$\Delta \lambda = \frac{\lambda_r^2}{2\pi R n_g} \qquad (2.42)$$

The nonlinearity of the fiber ring is of the Kerr-type, wherein the nonlinear refractive index is given by [42–44]

$$n = n_0 + n_2 I = n_0 + (\frac{n_2}{A_{eff}})P, \qquad (2.43)$$

where n_0 and n_2 are the linear and nonlinear refractive indices [45, 46], while I and P are the optical intensity and optical field power, respectively [47, 48]. Here, the fiber coupler is considered as a point device and is reciprocal. The linear and nonlinear phase shifts of the ring resonator can be expressed by $\phi_0 = kLn_0$ and $\phi_{NL} = kLn_2|E_1|^2$, where $k = 2\pi/\lambda$ is a wave number, and $L = 2\pi R$ is the circumference of the ring resonator, where R is the radius of the ring resonator [49–52]. Mathematically, the subsequence equations of the round-trip within the system is given by

$$E_{n+1} = j\sqrt{(1-\gamma)\kappa}E_{in} + \sqrt{(1-\gamma)(1-\kappa)}x E_n \exp(-j(\phi_0 + \phi_{NL})). \quad (2.44)$$

Here, the subscript n denotes the number of round-trips inside the system. This equation has to be satisfied with boundary conditions appropriate for ring. The transmission around the single ring resonator is represented by

$$z^{-1} = \exp(-\alpha L/2 - jk_n L) \qquad (2.45)$$

where k_n is the propagation constant and $\alpha L/2$ is the ring loss (round-trip loss), which includes propagation loss, losses resulting from transitions in the curvature, and bending losses [53, 54]. The value of α (unit length^{-1}) depends on the properties of the material and the waveguide used, and it is referred to as the intensity attenuation coefficient, where L is the circumference of the ring resonator. In order to define this, we consider a ring resonator connected to a single coupler that extracts light from the ring into the output waveguides.

When an input electric field, E_i is coupled to the ring waveguide through an external bus waveguide, a positive feedback is induced and the field inside the ring resonator, E_{r2} starts to build up. The feedback mechanism will be induced by the ring waveguide, therefore does not need any further requirements such as Bragg gratings, mirrors, or distributed feedback waveguides with difficult fabrication process. Due to on-resonant certain wavelength of the input signals inside the ring waveguide, frequency selectivity is obtained [55, 56]. The inserted and transmitted electric fields into the ring resonator are expressed by

$$E_{r1} = (1-\gamma)^{\frac{1}{2}} \left[jE_i\sqrt{\kappa} + E_{r2}\sqrt{1-\kappa} \right] \qquad (2.46)$$

$$E_{r2} = E_{r1} \exp(-\frac{\alpha}{2}L - jk_n L) \qquad (2.47)$$

where $k_n = \frac{2\pi \cdot n_{eff}}{\lambda}$ and γ denotes the intensity insertion loss coefficient of the directional coupler and n_{eff} is the effective refractive index [57, 58]. Therefore, the refractive index n quantifies the increase in the wavenumber (phase change per unit length) caused by the medium. Here, the effective refractive index n_{eff} has the similar meaning with light propagation in a waveguide, where it depends not only on the wavelength but also on the mode, in which the light propagates. The ratio of the output and input powers which is E_t/E_i can be calculated as [59–61]

$$\frac{E_t}{E_i} = (1 - \gamma)^{\frac{1}{2}} \cdot \left[\frac{\sqrt{1 - \kappa} - (1 - \gamma)^{\frac{1}{2}} \cdot \exp(-\frac{\alpha}{2}L - jk_nL)}{1 - (1 - \gamma)^{\frac{1}{2}} \cdot \sqrt{1 - \kappa} \cdot \exp(-\frac{\alpha}{2}L - jk_nL)} \right] \qquad (2.48)$$

In the following new parameter will be used for simplifying [62, 63]:

$$D = (1 - \gamma)^{\frac{1}{2}}, \quad x = D \cdot \exp(-\frac{\alpha}{2} \cdot L), \quad y = \sqrt{1 - \kappa}, \quad \phi = k_nL$$

Intensity relation to the output port is given by [64]:

$$T = \frac{I_t}{I_i}(\varphi) = \left| \frac{E_t}{E_i} \right|^2 = D^2 \cdot \left[1 - \frac{(1 - x^2) \cdot (1 - y^2)}{(1 - xy)^2 + 4xy \cdot \sin^2(\frac{\varphi}{2})} \right] \qquad (2.49)$$

Maximum and minimum transmission can be calculated when $\sin^2(\frac{\varphi}{2})$ is "1" and "0" respectively. Therefore;

$$T_{\max} = D^2 \cdot \frac{(x + y)^2}{(1 + x \cdot y)^2} \qquad (2.50)$$

$$T_{\min} = D^2 \cdot \frac{(x - y)^2}{(1 - x \cdot y)^2} \qquad (2.51)$$

The minimum transmission, $T_{\min}$ occurs at the resonant point when the circumference of the ring L, is an integer number of the guide wavelength, which is given by

$$\phi = k_n \cdot L = 2m\pi, \quad m = \text{integer}, \\ m \cdot \lambda_m = n \cdot L \qquad (2.52)$$

here, m is the mode number, λ_m is the resonant mode wavelength. The on-off ratio for the single ring resonator is defined as the ratio of the on-resonance intensity to the off-resonance intensity which is maximum at $T_{\min} = 0$. Therefore $x = y$ and

$$\alpha = -\frac{1}{L} \times \ln(\frac{1 - \kappa}{D^2}) \qquad (2.53)$$

This relationship given by Eq. (2.26) is also referred to as critical coupling, where the maximum on-off ratio $\frac{I_t}{I_i}(2m\pi) = 0$ can be obtained by varying the coupling coefficient (κ) or the intensity attenuation coefficient (α).

Fig. 2.3 Ring resonator with two adjacent waveguide

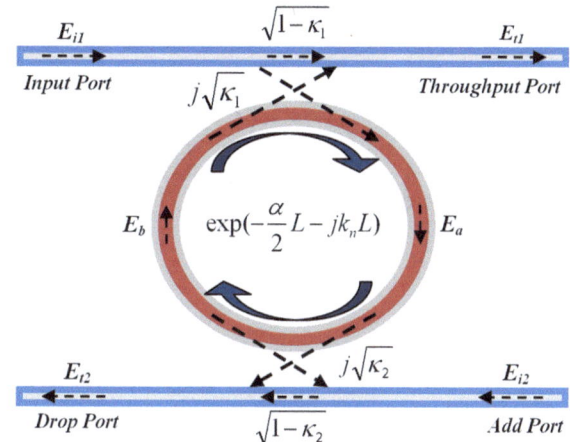

2.17 Ring Resonator as Add/Drop Filter

Recently, optical ring resonators (ORR) have numerous applications in single mode lasers, biosensors, optical switching, add/drop filters, tunable lasers, signal processing and dispersion compensators [65–67]. In any WDM system, optical filters are used for separating one optical channel from the combined signals. The basic ORR with two couplers is illustrated in Fig. 2.3. The main performance characteristics of these resonators are the transmittance, free spectral range, finesse, Q-factor, and the group delay, which have been demonstrated both theoretically and experimentally in many works. Structural design of a single ring resonator (SRR) add/drop filter system is shown in Fig. 2.3, which is constructed by 2×2 optical couplers.

For simplification, the intensity relation [68] does not take into account coupling loses ($D^2 = 1$).

$$E_a = E_{i1} j \sqrt{\kappa_1} + E_b \sqrt{1 - \kappa_1} e^{\frac{-\alpha}{2} \frac{L}{2} - jkn\frac{L}{2}} \tag{2.54}$$

$$E_b = E_a \sqrt{1 - \kappa_2} e^{\frac{-\alpha}{2} \frac{L}{2} - jk_n \frac{L}{2}} \tag{2.55}$$

$$E_a = \frac{E_{i1} j \sqrt{\kappa_1}}{1 - \sqrt{1 - \kappa_1}\sqrt{1 - \kappa_2} e^{\frac{-\alpha}{2} L - jk_n L}} \tag{2.56}$$

$$E_b = \frac{E_{i1} j \sqrt{\kappa_1}}{1 - \sqrt{1 - \kappa_1}\sqrt{1 - \kappa_2} e^{\frac{-\alpha}{2} L - jk_n L}} \cdot \sqrt{1 - \kappa_2} e^{\frac{-\alpha}{2} \frac{L}{2} - jk_n \frac{L}{2}} \tag{2.57}$$

$$E_{t1} = E_b e^{\frac{-\alpha}{2} \frac{L}{2} - jk_n \frac{L}{2}} j \sqrt{\kappa_1} + E_{i1} \sqrt{1 - \kappa_1} \tag{2.58}$$

$$E_{t2} = E_a e^{-\frac{\alpha}{2}\frac{L}{2}-jk_n\frac{L}{2}} j\sqrt{\kappa_2} \quad at \;\; E_{i2} = 0 \tag{2.59}$$

$$
\begin{aligned}
\frac{E_{t1}}{E_{i1}} &= \frac{-\kappa_1\sqrt{1-\kappa_2}e^{-\frac{\alpha}{2}L-jk_nL} + \sqrt{1-\kappa_1} - (1-\kappa_1)\sqrt{1-\kappa_2}e^{-\frac{\alpha}{2}L-jk_nL}}{1 - \sqrt{1-\kappa_1}\sqrt{1-\kappa_2}e^{-\frac{\alpha}{2}L-jk_nL}} \\
&= \frac{-\sqrt{1-\kappa_2}e^{-\frac{\alpha}{2}L-jk_nL} + \sqrt{1-\kappa_1}}{1 - \sqrt{1-\kappa_1}\sqrt{1-\kappa_2}e^{-\frac{\alpha}{2}L-jk_nL}}
\end{aligned}
\tag{2.60}
$$

$$\frac{E_{t2}}{E_{i1}} = \frac{-\sqrt{\kappa_1.\kappa_2}e^{-\frac{\alpha}{2}\frac{L}{2}-jk_n\frac{L}{2}}}{1 - \sqrt{1-\kappa_1}\sqrt{1-\kappa_2}e^{-\frac{\alpha}{2}L-jk_nL}} \tag{2.61}$$

where κ_1 and κ_2 are the coupling coefficients, $L = 2\pi R$ and R is the radius of the add/drop filter device [69, 70]. The normalized outputs of the add/drop filter system are expressed as [71, 72]:

$$\frac{I_{t1}}{I_{i1}} = \left|\frac{E_{t1}}{E_{i1}}\right|^2 = \frac{1 - \kappa_1 - 2\sqrt{1-\kappa_1}\sqrt{1-\kappa_2}e^{-\frac{\alpha}{2}L}\cos(k_nL) + (1-\kappa_2)e^{-\alpha L}}{1 + (1-\kappa_1)(1-\kappa_2)e^{-\alpha L} - 2\sqrt{1-\kappa_1}\sqrt{1-\kappa_2}e^{-\frac{\alpha}{2}L}\cos(k_nL)} \tag{2.62}$$

$$\frac{I_{t2}}{I_{i1}} = \left|\frac{E_{t2}}{E_{i1}}\right|^2 = \frac{\kappa_1 \cdot \kappa_2 e^{-\frac{\alpha}{2}L}}{1 + (1-\kappa_1)(1-\kappa_2)e^{-\alpha L} - 2\sqrt{1-\kappa_1}\sqrt{1-\kappa_2}e^{-\frac{\alpha}{2}L}\cos(k_nL)} \tag{2.63}$$

Using $y_1 = \sqrt{1-\kappa_1}$ and $y_2 = \sqrt{1-\kappa_2}$, the intensity relations are then given by:

$$\frac{I_{t1}}{I_{i1}}(\varphi) = \left|\frac{E_{t1}}{E_{i1}}\right|^2 = 1 - \frac{(1-y_1^2)\cdot(1-y_2^2x^2)}{(1-y_1y_2x)^2 + 4y_1y_2x\sin^2\left(\frac{\varphi}{2}\right)} \tag{2.64}$$

$$\frac{I_{t2}}{I_{i1}}(\varphi) = \left|\frac{E_{t2}}{E_{i1}}\right|^2 = \frac{(1-y_1^2)\cdot(1-y_2^2)\cdot x}{(1-y_1y_2x)^2 + 4y_1y_2x\sin^2\left(\frac{\varphi}{2}\right)} \tag{2.65}$$

The full-width at half-maximum (FWHM) is given in this configuration by:

$$\delta\phi = 2\frac{1-y_1y_2x}{\sqrt{y_1y_2x}}, \tag{2.66}$$

where the finesse F is given by:

$$F = \frac{2\pi}{\delta\phi} = \frac{\pi\sqrt{y_1y_2x}}{1-y_1y_2x} \tag{2.67}$$

The maximum and minimum transmission are calculated as follows. For the throughput port:

$$T_{\text{max}} = \frac{(y_1 + y_2 x)^2}{(1 + y_1 y_2 x)^2} \tag{2.68}$$

$$T_{\text{min}} = \frac{(y_1 - y_2 x)^2}{(1 - y_1 y_2 x)^2} \tag{2.69}$$

And for the drop port:

$$T_{\text{max}} = \frac{(1 - y_1^2) \cdot (1 - y_2^2) \cdot x}{(1 - y_1 y_2 x)^2} \tag{2.70}$$

$$T_{\text{min}} = \frac{(1 - y_1^2) \cdot (1 - y_2^2) \cdot x}{(1 + y_1 y_2 x)^2} \tag{2.71}$$

The on-off ratio of an add/drop filter system is given by:

$$\frac{T_{\text{max}}(through\,put\,port)}{T_{\text{min}}(drop\,port)} = \frac{(y_1 + y_2 x)^2}{(1 - y_1^2) \cdot (1 - y_2^2) \cdot x} \tag{2.72}$$

The output intensity, I_{t1} at the throughput port will be zero at resonance ($k_n L = 2m\pi$) which indicates that the resonance wavelength is fully extracted by the resonator when $\kappa_1 = \kappa_2$ and $\alpha = 0$. The loss of signal power resulting from the insertion of a device in a transmission line for example an optical fiber is defined insertion loss and usually expressed in *dBs*. Therefore, it is a measure of attenuation. Attenuation can include loss due to the source and load impedances not matching, but is not included in insertion loss since this is a loss that was already present before the "insertion" was made. If the power transmitted to the load before insertion is P_T and the power received by the load after insertion is P_R, then the insertion loss in dB is given by,

$$IL = 10 \log_{10} \frac{P_T}{P_R} \tag{2.73}$$

2.18 Ring Resonator as PANDA

This system consists of one add/drop interferometer system connected to two ring resonators in the left and right sides. This system represents a new technique of combination and integration of micro ring resonators in which it can be widely used to improve the secure communication and the high capacity of optical signal proceeding in network communications [73–75]. Here the derived equations of the system is introduced which show that how does the input pulse propagates inside the rings systems [76–78]. The proposed system is shown in Fig. 2.4.

The resonator output fields, E_{t1} and E_1 consist of the transmitted and circulated components within the add/drop optical filter system, given by [79–81]

Fig. 2.4 Schematic of a
PANDA ring resonator
system

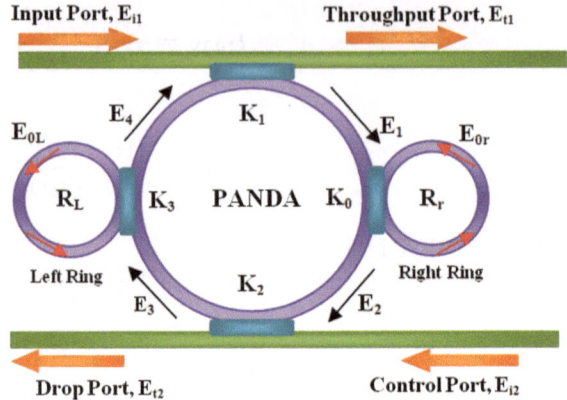

$$E_{t1} = \sqrt{1-\gamma_1}\left[\sqrt{1-\kappa_1}E_{i1} + j\sqrt{\kappa_1}E_4\right],\qquad(2.74)$$

$$E_1 = \sqrt{1-\gamma_1}\left[\sqrt{1-\kappa_1}E_4 + j\sqrt{\kappa_1}E_{i1}\right],\qquad(2.75)$$

$$E_2 = E_{0r}E_1 e^{-\frac{\alpha}{2}\frac{L}{2}-jk_n\frac{L}{2}},\qquad(2.76)$$

where κ_1 is the intensity coupling coefficient, γ_1 is the fractional coupler intensity
loss, α is the attenuation coefficient, $k_n = \frac{2\pi}{\lambda}$ is the wave propagation number, λ is
the input wavelength light field, $L = 2\pi R_{ad}$ and R_{ad} is the radius of the add/drop
system. For the second coupler of the add/drop system [82, 83],

$$E_{t2} = \sqrt{1-\gamma_2}\left[\sqrt{1-\kappa_2}E_{i2} + j\sqrt{\kappa_2}E_2\right],\qquad(2.77)$$

$$E_3 = \sqrt{1-\gamma_2}\left[\sqrt{1-\kappa_2}E_2 + j\sqrt{\kappa_2}E_{i2}\right],\qquad(2.78)$$

$$E_4 = E_{0L}E_3 e^{-\frac{\alpha}{2}\frac{L}{2}-jk_n\frac{L}{2}}.\qquad(2.79)$$

E_{0r} and E_{0L} are the light fields circulated components of the nanoring radii and R_r
and R_L are the coupled rings into the right and left sides of the add/drop optical filter system, respectively. Transmitted and circulated components of the light fields
in the right nanoring, R_r are given by

$$E_2 = \sqrt{1-\gamma}\left[\sqrt{1-\kappa_0}E_1 + j\sqrt{\kappa_0}E_{r2}\right],\qquad(2.80)$$

$$E_{r1} = \sqrt{1-\gamma}\left[\sqrt{1-\kappa_0}E_{r2} + j\sqrt{\kappa_0}E_1\right],\qquad(2.81)$$

$$E_{r2} = E_{r1}e^{-\frac{\alpha}{2}L_1 - jk_n L_1}.$$

(2.82)

or

$$E_{r1} = \frac{j\sqrt{1-\gamma}\sqrt{\kappa_0}\,E_1}{1 - \sqrt{1-\gamma}\sqrt{1-\kappa_0}\,e^{-\frac{\alpha}{2}L_1 - jk_n L_1}},$$

(2.83)

$$E_{r2} = \frac{j\sqrt{1-\gamma}\sqrt{\kappa_0}\,E_1 e^{-\frac{\alpha}{2}L_1 - jk_n L_1}}{1 - \sqrt{1-\gamma}\sqrt{1-\kappa_0}\,e^{-\frac{\alpha}{2}L_1 - jk_n L_1}},$$

(2.84)

where $L_1 = 2\pi R_r$ and R_r is the radius of the right side nanoring. Thus, the output circulated light field, E_{0r}, for the right side nanoring is given by [84, 85]

$$E_{0r} = E_1 \frac{\sqrt{(1-\gamma)(1-\kappa_0)} - (1-\gamma)e^{-\frac{\alpha}{2}L_1 - jk_n L_1}}{1 - \sqrt{1-\gamma}\sqrt{1-\kappa_0}\,e^{-\frac{\alpha}{2}L_1 - jk_n L_1}}.$$

(2.85)

Similarly, the output circulated light field, E_{0L}, for the left side nanoring of the add/drop system is given by

$$E_{0L} = E_3 \frac{\sqrt{(1-\gamma_3)(1-\kappa_3)} - (1-\gamma_3)e^{-\frac{\alpha}{2}L_2 - jk_n L_2}}{1 - \sqrt{1-\gamma_3}\sqrt{1-\kappa_3}\,e^{-\frac{\alpha}{2}L_2 - jk_n L_2}},$$

(2.86)

where $L_2 = 2\pi R_L$ and R_L is the radius of the left side nanoring. Regarding to more simplification such as $x_1 = (1-\gamma_1)^{1/2}$, $x_2 = (1-\gamma_2)^{1/2}$, $y_1 = (1-\kappa_1)^{1/2}$, and $y_2 = (1-\kappa_2)^{1/2}$, the interior circulated light fields, E_1, E_3 and E_4 are given by

$$E_1 = \frac{jx_1\sqrt{\kappa_1}E_{i1} + jx_1 x_2 y_1 \sqrt{\kappa_2}E_{0L}E_{i2}e^{-\frac{\alpha}{2}\frac{L}{2} - jk_n\frac{L}{2}}}{1 - x_1 x_2 y_1 y_2 E_{0r} E_{0L} e^{-\frac{\alpha}{2}L - jk_n L}},$$

(2.87)

$$E_3 = x_2 y_2 E_{0r} E_1 e^{-\frac{\alpha}{2}\frac{L}{2} - jk_n\frac{L}{2}} + jx_2\sqrt{\kappa_2}E_{i2},$$

(2.88)

$$E_4 = x_2 y_2 E_{0r} E_{0L} E_1 e^{-\frac{\alpha}{2}L - jk_n L} + jx_2\sqrt{\kappa_2}E_{0L}E_{i2}e^{-\frac{\alpha}{2}\frac{L}{2} - jk_n\frac{L}{2}}.$$

(2.89)

Thus, the throughput port (E_{t1}) output is expressed by

$$E_{t1} = AE_{i1} - BE_{i2}e^{-\frac{\alpha}{2}\frac{L}{2} - jk_n\frac{L}{2}}\left[\frac{CE_{i1}\left(e^{-\frac{\alpha}{2}\frac{L}{2} - jk_n\frac{L}{2}}\right)^2 + DE_{i2}\left(e^{-\frac{\alpha}{2}\frac{L}{2} - jk_n\frac{L}{2}}\right)^3}{1 - F\left(e^{-\frac{\alpha}{2}\frac{L}{2} - jk_n\frac{L}{2}}\right)^2} \right]$$

(2.90)

where, $A = x_1 x_2$, $B = x_1 x_2 y_2 \sqrt{\kappa_1}E_{0L}$, $C = x_1^2 x_2 k_1 \sqrt{\kappa_2}E_{0r}E_{0L}$, $D = (x_1 x_2)^2 y_1 y_2 \sqrt{\kappa_1 \kappa_2}E_{0r}E_{0L}^2$ and $F = x_1 x_2 y_1 y_2 E_{0r}E_{0L}$. The power output of the throughput port (P_{t1}) is given by

$$P_{t1} = (E_{t1}) \cdot (E_{t1})^* = |E_{t1}|^2 \tag{2.91}$$

Similarly, the output optical field of the drop port (E_{t2}) is given by,

$$E_{t2} = x_2 y_2 E_{i2} \left[\frac{x_1 x_2 \sqrt{\kappa_1 \kappa_2} E_{0r} E_{i1} e^{-\frac{\alpha}{2} \frac{L}{2} - jk_n \frac{L}{2}} + x_1 x_2^2 y_1 y_2 \sqrt{\kappa_2} E_{0r} E_{0L} E_{i2} \left(e^{-\frac{\alpha}{2} \frac{L}{2} - jk_n \frac{L}{2}} \right)^2}{1 - x_1 x_2 y_1 y_2 E_{0r} E_{0L} \left(e^{-\frac{\alpha}{2} \frac{L}{2} - jk_n \frac{L}{2}} \right)^2} \right],$$

$$\tag{2.92}$$

where the power output of the drop port (P_{t2}) is expressed by

$$P_{t2} = (E_{t2}) \cdot (E_{t2})^* = |E_{t2}|^2 \tag{2.93}$$

References

1. Amiri IS, Alavi SE, Idrus SM (2014) Soliton coding for secured optical communication link. Springer, USA
2. Amiri IS, Gifany D, Ali J (2013) Ultra-short multi soliton generation for application in long distance communication. J Basic Appl Sci Res (JBASR) 3(3):442–451
3. Meijerink A (2005) Coherence multiplexing for optical communication systems. University of Twente, The Netherlands
4. Recommendation I (2006) Spectral grids for WDM applications: DWDM frequency grid. In. G
5. Benson TM, Boriskina SV, Sewell P, Vukovic A, Greedy SC, Nosich AI (2006) Micro-optical resonators for microlasers and integrated optoelectronics. Frontiers Planar Lightwave Circ Technol 216:39–70
6. Yariv A, Xu Y, Lee RK, Scherer A (1999) Coupled-resonator optical waveguide: a proposal and analysis. Opt Lett 24(11):711–713
7. Singh SP, Singh N (2007) Nonlinear effect in optical fibers; origin, management and applications. Prog Electromagnet Res PIER 73:249–275
8. Kärtner F, Keller U (1995) Stabilization of soliton like pulses with a slow saturable absorber. Opt Lett 20(1):16–18
9. Marchese SV, Südmeyer T, Golling M, Grange R, Keller U (2006) Pulse energy scaling to 5 μJ from a femtosecond thin disk laser. Opt Lett 31(18):2728–2730
10. Anderson D, Desaix M, Karlsson M, Lisak M, Quiroga-Teixeiro M (1993) Wave-breaking-free pulses in nonlinear-optical fibers. JOSA B 10(7):1185–1190
11. Tomlinson W (1989) Curious features of nonlinear pulse propagation in single-mode optical fibers. Optics News 15(1):7–11
12. Fyath R, Al-mfrji AA (2012) Investigation of chaos synchronization in photonic crystal lasers. Opt Laser Technol 44:1406–1419
13. Yang L, Pan W, Yan LS, Luo B, Xiang SY, Jiang N (2011) Study on the dynamic and unpredictability properties of an optical microring resonator with modulation. Opt Eng 50:084203
14. Amiri IS, Jalil MA, Afroozeh A, Kouhnavard M, Ali J, Yupapin PP (2010) Controlling center wavelength and free spectrum range by MRR Radii. In: Faculty of science postgraduate conference (FSPGC), Universiti Teknologi Malaysia, 5–7 Oct 2010
15. Amiri IS, Nikmaram M, Shahidinejad A, Ali J (2012) Cryptography scheme of an optical switching system using pico/femto second soliton pulse. Int J Adv Eng Technol (IJAET) 5(1):176–184
16. Amiri IS, Nikoukar A, Ali J (2013) GHz frequency band soliton generation using integrated ring resonator for WiMAX optical communication. Opt Quantum Electron. doi:10.1007/s11082-013-9848-0

17. Kennedy MP, Kolumbán G (2000) Digital communications using chaos. Sign Proces 80(7):1307–1320
18. Kuleshov V, Larionova M, Udalov N (1998) Information transmission system with chaotic carrier: demands to the accuracy of symbols generation. In: Proceedings of the 1998 international symposium on acoustoelectronics, frequency control and signal generation, St.-Petersburg, Russia, 192–199
19. Kouhnavard M, Amiri IS, Jalil M, Afroozeh A, Ali J, Yupapin PP (2010) QKD via a quantum wavelength router using spatial soliton. AIP Conf Proc 1347:210–216
20. Amiri IS, Nikoukar A, Ali J (2010) Quantum information generation using optical potential well. Paper presented at the network technologies and Communications (NTC) conference, Singapore
21. Freysz E, Degert J (2010) Nonlinear optics: Terahertz Kerr effect. Nat Photonics 4(3):131–132
22. Agrawal GP (1999) Fiber-optic communication systems 1997, vol 6. Wiley, New York, pp 1093–1102
23. Agrawal G (1995) Nonlinear fiber optics. In Academic Press, San Diego
24. Mitchell M (1996) An introduction to genetic algorithms, 1996. PHI Pvt. Ltd., New Delhi
25. Kivshar YS, Pelinovsky DE (2000) Self-focusing and transverse instabilities of solitary waves. Phys Rep 331(4):117–195
26. Zakharov VE, Shabat AB (1972) Exact theory of two-dimensional self-focusing and one-dimensional self-modulation of waves in nonlinear media (Differential equation solution for plane self focusing and one dimensional self modulation of waves interacting in nonlinear media). Soviet Physics-JETP 34:62–69
27. Nakazawa M, Kubota H, Suzuki K, Yamada E, Sahara A (2000) Ultrahigh-speed long-distance TDM and WDM soliton transmission technologies. Sel Top Quantum Electron, IEEE J 6(2):363–396
28. Dell'Anno F, De Siena S, Illuminati F (2006) Multiphoton quantum optics and quantum state engineering. Phys rep 428(2), 53–168
29. Kivshar YS, Luther-Davies B (1998) Dark optical solitons: physics and applications. Phys Rep 298(2):81–197
30. Honarasa G, Hatami M, Tavassoly M (2011) Quantum Squeezing of dark solitons in optical fibers. Commun Theor Phys 56:322
31. Sheng-Mei A, Jia-Ren Y (2006) Effect of higher-order terms on nonlinear Schrödinger dark solitons in optical fibres. Chin Phys Lett 23:2774
32. Mitatha S, Chaiyasoonthorn N, Yupapin PP (2009) Dark-bright optical solitons conversion via an optical add/drop filter. Microwave Opt Technol Lett 51(9):2104–2107
33. Infeld E, Rowlands G (2000) Nonlinear waves, solitons, and chaos. Cambridge University Press, Cambridge
34. Liang W, Zhi-Jie G, Li-Jun S (2010) Properties of pulsating solitons in dissipative systems. Chin Phys B 19:080512
35. Yariv A (1991) Optical electronics (Saunders, Philadelphia, Pa. Author Affiliations Rabi Rabady, Ivan Avrutsky. Department of Electrical and Computer Engineering, Wayne State University, Detroit, Michigan, 48202)
36. Afroozeh A, Bahadoran M, Amiri IS, Samavati AR, Ali J, Yupapin PP (2012) Fast light generation using GaAlAs/GaAs waveguide. Jurnal Teknologi (Sci Eng) 57:17–23
37. Amiri IS, Nikoukar A, Ali J (2013) New system of chaotic signal generation based on coupling coefficients applied to an add/drop system. Int J Adv Eng Technol (IJAET) 6(1):78–87
38. Nikoukar A, Amiri IS, Ali J (2011) Secured binary codes generation for computer network communication. Paper presented at the network technologies and communications (NTC) conference, Singapore
39. Shahidinejad Ali (2014) Iraj Sadegh Amiri, Toni Anwar: enhancement of indoor WDM-based optical wireless communication using microring resonator. Rev Theor Sci 2(3):201–210
40. Amiri IS, Ali J (2013) Nano particle trapping by ultra-short tweezer and wells using MRR interferometer system for spectroscopy application. Nanosci Nanotechnol Lett 5(8):850–856

41. Amiri IS, Barati B, Sanati P, Hosseinnia A, Mansouri Khosravi HR, Pourmehdi S, Emami A, Ali J (2013) Optical stretcher of biological cells using sub-nanometer optical tweezers generated by an add/drop MRR system. Nanosci Nanotechnol Lett 6(2):111–117. doi:10.1166/nnl.2013.1738

42. Afroozeh A, Amiri IS, Ali J, Yupapin PP (2012) Determination of FWHM for soliton trapping. Jurnal Teknologi (Sci Eng) 55:77–83

43. Sanati P, Afroozeh A, Amiri IS, Ali J, Chua LS (2013) Femtosecond pulse generation using microring resonators for eye nano surgery. Nanosci Nanotechnol Lett 6:904–910

44. Alavi SE, Amiri IS, Idrus SM, Supa'at ASM (2014) Optical amplification of tweezers and bright soliton using an interferometer ring resonator system. J Comput Theor Nanosci (CTN)

45. Amiri IS, Nikoukar A, Shahidinejad A, Ranjbar M, Ali J, Yupapin PP (2012) Generation of quantum photon information using extremely narrow optical tweezers for computer network communication. GSTF J Comput 2(1):140

46. Zeinalinezhad A, Pourmand SE, Amiri IS, Afroozeh A (2014) Stop light generation using nano ring resonators for ROM. J Comput Theor Nanosci (CTN)

47. Amiri IS, Shahidinejad A, Nikoukar A, Ranjbar M, Ali J, Yupapin PP (2012) Digital binary codes transmission via TDMA networks communication system using dark and bright optical soliton. GSTF J Comput (joc) 2(1):12

48. Nikoukar A, Amiri IS, Ali J (2013) Generation of nanometer optical tweezers used for optical communication networks. Int J Innovative Res Comput Commun Eng 1(1):77–85

49. Afroozeh A, Kouhnavard M, Amiri IS, Jalil MA, Ali J, Yupapin PP (2010) Effect of center wavelength on MRR performance. In: Faculty of science postgraduate conference (FSPGC), Universiti Teknologi Malaysia, 5–7 Oct 2010

50. Amiri IS, Shahidinejad A, Ali J (2014) Generating of 57–61 GHz frequency band using a panda ring resonator Quantum Matter

51. Amiri IS, Afroozeh A, Bahadoran M, Ali J, Yupapin PP (2012) Molecular transporter system for qubits generation. Jurnal Teknologi (Sci Eng) 55:155–165

52. Amiri IS, Ali J (2014) Simulation of the single ring resonator based on the Z-transform method theory. Quantum Matter 3(6). doi:10.1166/qm.2014.1157

53. Amiri IS, Ali J (2014) Femtosecond optical quantum memory generation using optical bright soliton. J Comput Theor Nanosci (CTN) 11(6). doi:10.1166/jctn.2014.3521

54. Yupapin PP, Jalil MA, Amiri IS, Naim I, Ali J (2010) New communication bands generated by using a soliton pulse within a resonator system. Circ Syst 1(2):71–75

55. Glaser W (1997) Photonik für Ingenieure. Verl. Technik

56. Yupapin PP, Pornsuwancharoen N (2008) Guided wave optics and photonics: micro-ring resonator design for telephone network security. Nova Science Publishers, New York

57. Amiri IS, Vahedi G, Nikoukar A, Shojaei A, Ali J, Yupapin PP (2012) Decimal Convertor application for optical wireless communication by generating of dark and bright signals of soliton. Int J Eng Res Technol (IJERT) 1(5)

58. Jalil MA, Amiri IS, Kouhnavard M, Afroozeh A, Ali J, Yupapin PP (2010) Finesse improvements of light pulses within MRR system. In: Faculty of science postgraduate conference (FSPGC), Universiti Teknologi Malaysia, 5–7 Oct 2010

59. Amiri IS, Afroozeh A, Ali J, Yupapin PP (2012) Generation of quantum codes using up and down link optical soliton. Jurnal Teknologi (Sci Eng) 55:97–106

60. Amiri IS, Ali J (2014) Optical quantum generation and transmission of 57–61 GHz frequency band using an optical fiber optics. J Comput Theor Nanosci (CTN) 11(10). doi:10.1166/jctn.2014.3617

61. Shahidinejad A, Soltanmohammadi S, Amiri IS, Anwar T (2014) Solitonic pulse generation for inter-satellite optical wireless communication. Quantum Matter 3(2):150–154. doi:http://dx.doi.org/10.1166/qm.2014.1107

62. Alavi SE, Amiri IS, Supa'at ASM, Idrus SM (2014) Indoor data transmission over ubiquitous infrastructure of powerline cables and LED lighting. J Comput Theor Nanosci (CTN)

63. Amiri IS, Ali J (2013) Optical buffer application used for tissue surgery using direct interaction of nano optical tweezers with nano cells. Quantum Matter 2(6):484–488

64. Okamoto K (2006) Fundamentals of optical waveguides. Academic press, London
65. Amiri IS, Alavi SE, Idrus SM, Supa'at ASM, Ali J, Yupapin PP (2013) All optical OFDM generation for IEEE802.11a based on soliton carriers using microring resonators. IEEE Photonics J 6(1). doi:10.1109/JPHOT.2014.2302791
66. Amiri IS, Alavi SE, Bahadoran M, Afroozeh A, Idrus SM, Ali J (2014) Nanometer Bandwidth soliton generation and experimental transmission within nonlinear fiber optics using an add-drop filter system. J Comput Theor Nanosci (CTN)
67. Amiri IS, Naraei P (2014) Optical transmission characteristics of an optical add-drop interferometer system. Quantum Matter
68. Yariv A (2000) Universal relations for coupling of optical power between micro resonators and dielectric waveguides. Electron Lett 36(4):321–322
69. Amiri IS, Gifany D, Ali J (2013) Entangled photon encoding using trapping of picoseconds soliton pulse. IOSR J Appl Phys (IOSR-JAP) 3(1):25–31
70. Amiri IS, Nikoukar A, Shahidinejad A, Anwar T (2014) The proposal of high capacity GHz soliton carrier signals applied for wireless commutation. Rev Theor Sci 2(4). doi:10.1166/rits.2014.1027
71. Gifany D, Amiri IS, Ranjbar M, Ali J (2013) Logic codes generation and transmission using an encoding-decoding system. Int J Adv Eng Technol (IJAET) 5(2):37–45
72. Nikoukar A, Amiri IS, Shahidinejad A, Shojaei A, Ali J, Yupapin PP (2012) MRR quantum dense coding for optical wireless communication system using decimal convertor. In: Computer and communication engineering (ICCCE) conference, Malaysia 2012, IEEE Explore, pp 770–774
73. Amiri IS, Ali J (2014) Deform of biological human tissue using inserted force applied by optical tweezers generated by PANDA ring resonator. Quantum Matter 3(1):24–28
74. Amiri IS, Ali J, Yupapin PP (2012) Enhancement of FSR and Finesse using add/drop filter and PANDA ring resonator systems. Int J Mod Phys B 26(04):1250034
75. Amiri IS, Afroozeh A, Bahadoran M (2011) Simulation and analysis of multisoliton generation using a PANDA ring resonator system. Chin Phys Lett 28(10):104205
76. Reece P, Wright E, Dholakia K (2007) Experimental observation of modulation instability and optical spatial soliton arrays in soft condensed matter. Phys Rev Lett 98(20):203902
77. Brambilla G, Murugan GS, Wilkinson J, Richardson D (2007) Optical manipulation of microspheres along a subwavelength optical wire. Opt Lett 32(20):3041–3043
78. Shvedov VG, Rode AV, Izdebskaya YV, Desyatnikov AS, Krolikowski W, Kivshar YS (2010) Selective trapping of multiple particles by volume speckle field. Opt Express 18(3):3137–3142
79. Amiri IS, Alavi SE, Idrus SM, Nikoukar A, Ali J (2013) IEEE 802.15.3c WPAN standard using millimeter optical soliton pulse generated by a PANDA ring resonator. IEEE Photonics J 5(5):7901912. doi:10.1109/JPHOT.2013.2280341
80. Afroozeh A, Amiri IS, Samavati A, Ali J, Yupapin PP (2012) THz frequency generation using MRRs for THz imaging. In: International conference on enabling science and nanotechnology (EsciNano), Kuala Lumpur, Malaysia 2012, IEEE Explore, pp 1–2
81. Amiri IS, Ebrahimi M, Yazdavar AH, Gorbani S, Alavi SE, Idrus SM, Ali J (2013) Transmission of data with OFDM technique for communication networks using GHz frequency band soliton carrier. IET Communications
82. Amiri SI, Nikmaram M, Shahidinejad A, Ali J (2013) Generation of potential wells used for quantum codes transmission via a TDMA network communication system. Secur Commun Netw 6(11):1301–1309. doi:10.1002/sec.712
83. Amiri IS, Shahidinejad A, Nikoukar A, Ali J, Yupapin PP (2012) A Study of dynamic optical tweezers generation for communication networks. Int J Adv Eng Technol (IJAET) 4(2):38–45
84. Amiri IS, Ali J (2014) Multiplex and de-multiplex of generated multi optical soliton by MRRs using fiber optics transmission link. Quantum Matter
85. Amiri IS, Nikoukar A, Ali J, Yupapin PP (2012) Ultra-short of pico and femtosecond soliton laser pulse using microring resonator for cancer cells treatment. Quantum Matter 1(2):159–165

Chapter 3
Integrated Ring Resonator Systems

Abstract To generate a spectrum of light over a broad range, an optical soliton pulse is recommended as a powerful laser pulse, which can be used to generate chaotic filter characteristics when propagating within nonlinear micro ring resonators (NMRRs). Propagation of input soliton pulses or any type of laser pulses inside the ring system causes the signals to be split and sliced into many smaller continuous signals (chaotic) due to the Kerr effect of the nonlinear system. Chaotic signals propagate through the ring resonator system, where the multi bright and dark soliton can be generated at the through and drop ports of the ring system. Ring resonators can be used to generate ultra-short pulses in the nonlinear regime, where the use of soliton laser has become an interesting subject. In such a way, the security and capacity of the communication system can be improved and used for many applications in optical communication networks.

Keywords Iterative method · Soliton · Nonlinear microring resonators (MRR) · Chaotic signal · Dark Soliton · Bright soliton · Ultra-short Pulse · Soliton Entanglement

3.1 Technical Progress

An iterative method is a mathematical process that generates a sequence of improving approximate solutions for a class of problems. In the problems of a system of equations, an iterative method uses an initial guess to generate successive estimates to a solution. In contrast, direct method attempts to solve the problem by a finite sequence of operations, where this method would deliver an exact solution. Iterative method is the only option for nonlinear equations, where it is often useful even for linear problems involving a large number of variables. In this study, the nonlinear equations of pulses propagating within the ring resonator are solved and coded to be simulated via a MATLAM software using iterative method. In order to validate the simulated results with experimental studies, the parameters of the ring system have been selected from real and practical implemented works.

© The Author(s) 2015
I. Sadegh Amiri and A. Afroozeh, *Ring Resonator Systems to Perform Optical Communication Enhancement Using Soliton*, SpringerBriefs in Applied Sciences and Technology, DOI 10.1007/978-981-287-197-8_3

3.2 Noisy or Chaotic Signals Generation

To generate a spectrum of light over a broad range, an optical soliton pulse is recommended as a powerful laser pulse which can be used to generate chaotic filter characteristics when propagating within nonlinear micro ring resonators (NMRRs). Using this technique, the capacity of the transmission data can be secured and increased when the chaotic packet switching is employed.

Recently, a number of devices are employed to transmit information using chaotic signals. Information signal can be injected into the chaotic signal generator. Therefore, information can be added to the chaotic signal generating system at the output of the system, which is known as chaotic masking, or it can modulate one of the chaotic generator parameters which called chaotic switching. However, the transmitting signals of chaotic comprise the injected information, where it is hidden and secured.

At the end of the transmission link the receiver which possesses this information, will reproduce the original input signal, where the technique of extracting information components from the chaotic carrying signal can be performed. Therefore, the chaotic signal generator system can be used to handle three important functions simultaneously, wherein it carries an information message from the source to the receiver via a transmission link, hides and secures the content from strangers, and plays a role of a synchronizing signal in the receiving system [1].

The system of chaotic signal generation is shown in Fig. 3.1. A series of microring resonators can be integrated in one single ring system. Propagation of input soliton pulses or any types of laser pulses inside the ring system cases the signals to be split and sliced into many smaller continues signals due to the Kerr effect of the nonlinear system. Therefore chaotic output signals from each ring resonator can be generated, where the filtering progress can be done by adding second and third ring resonators to the system.

Input light of the monochromatic laser beam is introduced into the system. The fiber has a nonlinear refractive index of n_2 and a linear absorption coefficient of α.

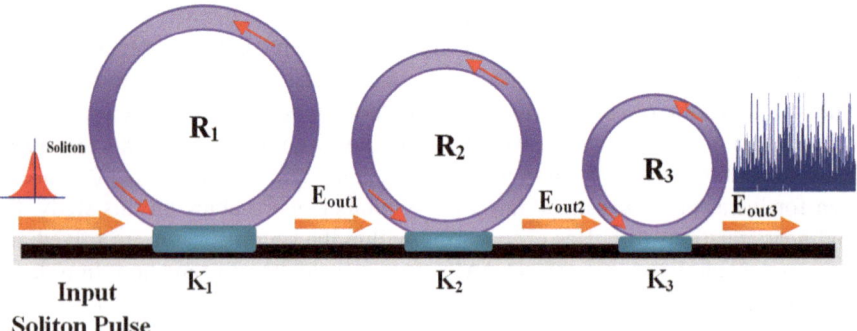

Fig. 3.1 Schematic of ring resonator system, where R_s ring radii, κ_s coupling coefficients, E_{out} output signals from each ring resonator

The intensity coupling coefficient of the fiber coupler is κ, where γ is a coupling loss of the field amplitude. The fiber ring has a resonant condition for the specific wavelength in the linear case. The E_{out} is representing output signal from each ring resonator. A_{eff} is the effective mode core area of the device [2], where for an add/drop optical filter design, the effective mode core areas range from 0.50 to 0.10 μm². The parameters were obtained by using practical parameters of used material (InGaAsP/InP). In this case input optical fields of bright and dark soliton are inserted into the ring system [3]. The methodology steps can be seen as following flowchart.

3.3 Conversion Technique to Form the Dark Soliton

The system of dark soliton generation consists of a series of ring resonators which are connected to an add/drop filter system shown in Fig. 3.2. To retrieve the signals from the chaotic noise, we propose to use the add/drop device with the appropriate parameters [4–12]. There are basically three essential parameters describing the behavior of a MRR filter response which are the –3 dB bandwidth or the full-width at half-maximum (FWHM), the on–off ratio, and the shape factor. For a lossless fiber ring resonator, the –3 dB bandwidth depends mainly on the coupling coefficients and the optical round-trip length. The on–off ratio for the throughput and drop ports, which is the ratio of the on-resonance intensity to the off-resonance intensity, is given by Eq. 2.47. The chaotic noise cancellation can be managed by using the specific parameters of the add/drop device, where the required signals can be retrieved by the specific users. To generate the dark

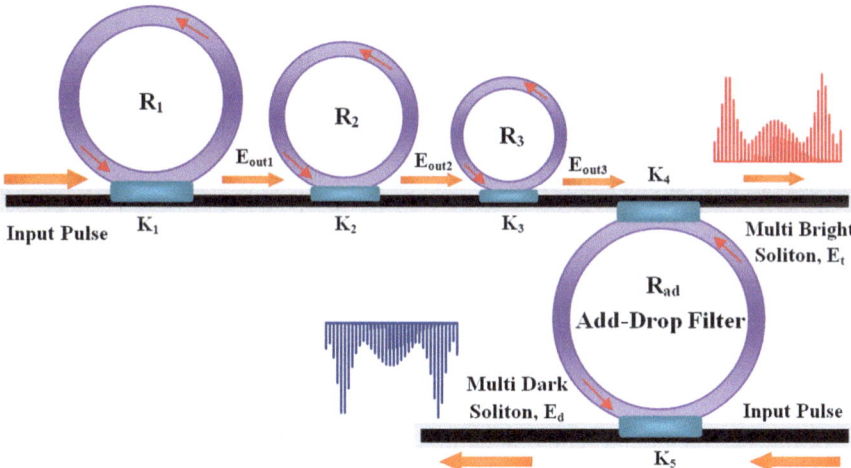

Fig. 3.2 Schematic of an optical dark soliton pulse generator system, R_s ring radii, κ_s coupling coefficients, κ_4 and κ_5 are the add/drop coupling coefficients, E_t and E_d through and drop port outputs respectively

soliton, optical field of bright soliton or Gaussian laser beam are introduced to the ring system [13–22].

Bright soliton in the form of chaotic signal output from the first ring with a variety of wavelengths passes through next ring resonators, where the filtering process occurs after each ring. Compress bandwidth with smaller group velocity will be achieved, where the second and third rings with lower radius are connected to the first ring. Therefore cancellation of some chaotic signals occurs, leading to generate high amplified pulses of bright soliton. Here the multi chaotic signals can be performed used for high security and capacity optical communication [23–31].

Chaotic signals propagate through the add/drop filter system, where the multi bright and dark soliton can be generated at the through and drop ports of the ring system. Generation of optical dark soliton performs the security applications of such as pulses in optical communication network due to the low power of their central wavelength, where they cannot be detected by most conventional optical detectors. The system of dark soliton generation can be used as an optical conversion system as well, where the dark soliton can be converted to a bright soliton and vice versa with suitable parameters of the rings.

The chaotic noise cancellation can be managed by using the specific parameters of the add-drop device in which required signals can be retrieved by the specific users. The waveguide (ring resonator) loss is $\alpha = 0.5$ dB mm^{-1}. The fractional coupler intensity loss is $\gamma = 0.1$. The methodology steps can be seen as following flowchart.

3.4 Highly Ultra-Short Pulse Generation

Ring resonators can be used to generate ultra-short pulses in the nonlinear regime, where the using of soliton laser becomes an interesting subject. High optical output signals of the ring system are of benefit for patterning thin-film. The system of dark soliton generation can be used to trap extremely short single and multisoliton by adding a fourth ring with small radius to the add/drop filter system and changing the variable parameters [32–40].

In such a way, the security and capacity of the ring system can be improved and used for many applications in optical communication networks. Therefore, adding the forth ring resonator to the ring system causes more filtering of the throughput output signals. The system of ultra-short pulse generation is shown in Fig. 3.3, while extremely short single and multiple optical soliton pulses in the range of nm/ns, pm/ps and fm/fs can be obtained using appropriate parameters of the system.

Localized spatial and temporal soliton pulses are useful to generate entangled photon pair that can provide continuous variable quantum key distribution (QKD) applicable for secured communication networks. The classical information and security code can be formed by using the temporal and spatial soliton pulses, respectively.

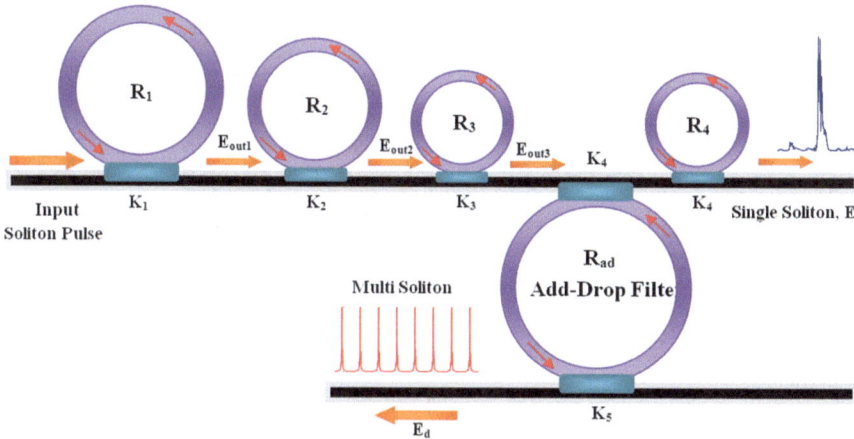

Fig. 3.3 A schematic of an extremely short single and multisoliton pulses generator, where R_s ring radii, κ_s coupling coefficients, κ_4 and κ_5 are add/drop coupling coefficients

3.5 Soliton Entanglement

The use of dark soliton array behavior has become a promising application, wherein the transmission of dark soliton can be converted into a bright soliton after passing through a specific add/drop filter. Therefore, the transmission signals can be converted into the form of dark solitons, which is difficult to detect, whereas the specific end user that connects to the link via the specific add/drop filter can detect and receive the signals. Experimentally, the dark soliton array was generated and detected by using the Brillouin Enhanced Fiber Laser (BEFL) scheme [41–47].

The system of nonlinear ring resonators can be used to generate multiple soliton channels. We propose a modified add/drop optical filter called PANDA system that consists of one centered ring resonator connected to one smaller ring resonators on the left side shown in Fig. 3.4.

Fig. 3.4 System of PANDA ring resonator, where (a): E_{i1} and E_{i2} are the input pulses, (b): E_{t1} and E_{t2} are the output signals, K is the coupling coefficient, E_1, E_2 and E_3 are the interior intensities inside the system

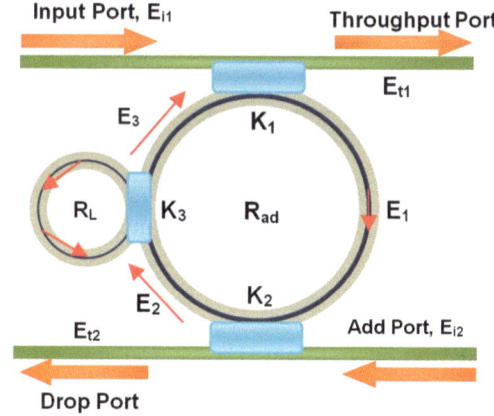

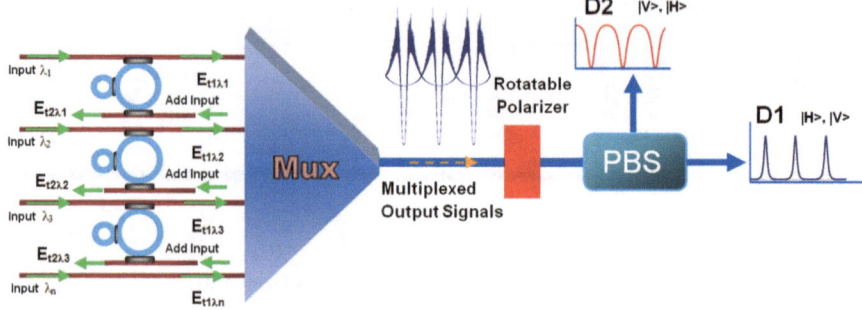

Fig. 3.5 System of integrated MRR systems, incorporating with a multiplexer device and a beam splitter

The input optical pulses will be inserted into the input and add port of the system, where the output signals will be seen in the throughput and drop ports. The interferometer process occurs inside the ring system leading to generate ultra-short of the dark and bright soliton. To form the multifunction operations of the PANDA system, for instance, to control, tune, and amplify, additional input pulse is introduced into the add port of the system. By controlling some suitable parameters of add optical pulse, the generated result within the ring resonator system can be controlled. In order to achieve high capacity of a transmission, output signals with different central wavelengths can be combined using a suitable multiplexer device. Here, a series of MRRs can be integrated in one single system, incorporating with multiplexer device shown in Fig. 3.5 [48–59].

Transmitted signals of dark soliton pulses from multiplexer system pass through a PBS device. The used beam splitter reflects (and transmit) 50 % of the light that is incident, for all polarizations of the incident light. This interconnection can also be done with fiber couplers. Therefore, different central wavelengths of laser pulses can be combined together using proposed system, wherein the single dark or bright soliton pulses join together making array output pulses, which can be used to increase the capacity of the ring system. The methodology steps can be seen as following flowchart [60–69].

3.6 Multiple Soliton Generation

Generation of multisoliton becomes an interesting subject when it is used to enlarge the capacity of communication channels. The high optical output of the ring resonator system is of benefit to long distance communication links. Gaussian pulse can be used to form a multisoliton using a ring resonator. We propose a PANDA ring resonator system to generate highly chaotic signal connected to an add/drop filter system. Therefore, the PANDA ring resonator can be connected to an add/drop filter system in order to filter noisy and chaotic signals. The proposed

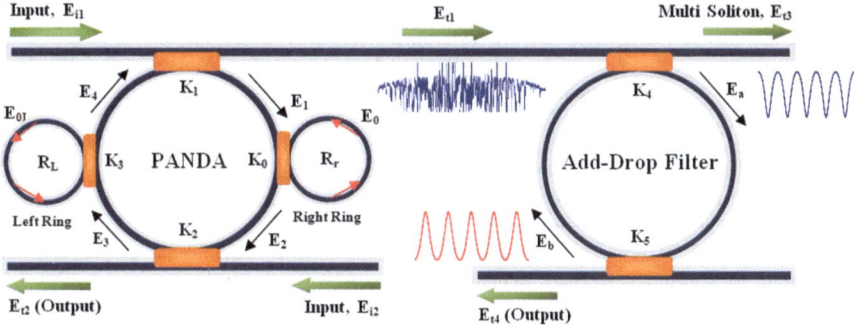

Fig. 3.6 System of multi dark and bright soliton generation

system consists of a PANDA ring resonator connected to an add/drop filter system, shown in Fig. 3.6 [70–84].

References

1. Pecora LM, Carroll TL (1990) Synchronization in chaotic systems. Phys Rev Lett 64(8):821–824
2. White T, McPhedran R, de Sterke CM, Botten L, Steel M (2001) Confinement losses in microstructured optical fibers. Opt Lett 26(21):1660–1662
3. Lin Y, Lee RK (2007) Dark-bright soliton pairs in nonlocal nonlinear media. Opt Express 15:8781–8786
4. Nikoukar A, Amiri IS, Alavi SE, Shahidinejad A, Anwar T, Supa'at ASM, Idrus SM, Teng LY (2014) Theoretical and simulation analysis of the add/drop filter ring resonator based on the Z-transform method theory. The 2014 third ICT International Student Project Conference (ICT-ISPC2014). IEEE, Thailand
5. Ali J, Afroozeh A, Amiri IS, Jalil MA, Yupapin PP (2010) Dark and bright soliton trapping using NMRR. ICEM. Legend Hotel, Kuala Lumpur, Malaysia
6. Ali J, Nur H, Lee S, Afroozeh A, Amiri IS, Jalil M, Mohamad A, Yupapin PP (2010) Short and millimeter optical soliton generation using dark and bright soliton. AMN-APLOC International Conference, Wuhan, China
7. Amiri IS, Afroozeh A, Nawi IN, Jalil MA, Mohamad A, Ali J, Yupapin PP (2011) Dark soliton array for communication security. Procedia Eng 8:417–422
8. Amiri IS, Shahidinejad A, Nikoukar A, Ranjbar M, Ali J, Yupapin PP (2012) Digital binary codes transmission via TDMA networks communication system using dark and bright optical soliton. GSTF J Comput 2(1):12
9. Amiri IS, Barati B, Sanati P, Hosseinnia A, Mansouri Khosravi HR, Pourmehdi S, Emami A, Ali J (2014) Optical stretcher of biological cells using sub-nanometer optical tweezers generated by an add/drop microring resonator system. Nanosci Nanotechnol Lett 6(2):111–117
10. Amiri IS, Ali J (2014) Generating highly dark–bright solitons by gaussian beam propagation in a PANDA ring resonator. J Comput Theor Nanosci 11(4):1092–1099
11. Amiri IS, Naraei P (2014) Optical transmission characteristics of an optical add-drop interferometer system. Quantum Matter
12. Alavi SE, Amiri IS, Idrus SM, Supa'at ASM, Ali J, Yupapin PP (2014) All optical OFDM generation for IEEE802.11a based on soliton carriers using microring resonators. IEEE Photonic J 6(1)

13. Afroozeh A, Bahadoran M, Amiri IS, Samavati AR, Ali J, Yupapin PP (2012) Fast light generation using GaAlAs/GaAs waveguide. Jurnal Teknologi 57:17–23
14. Ali J, Raman K, Kouhnavard M, Amiri IS, Jalil MA, Afroozeh A, Yupapin PP (2011) Dark soliton array for communication security. AMN-APLOC International Conference, Wuhan, China
15. Amiri IS, Afroozeh A, Bahadoran M (2011) Simulation and analysis of multisoliton generation using a PANDA ring resonator system. Chin Phys Lett 28(10):104–205
16. Amiri IS, Afroozeh A, Bahadoran M, Ali J, Yupapin PP (2012) Molecular transporter system for qubits generation. Jurnal Teknologi 55:155–165
17. Amiri IS, Ali J (2013) Data signal processing via a manchester coding-decoding method using chaotic signals generated by a PANDA ring resonator. Chin Opt Lett 11(4):041901–041904
18. Amiri IS, Nikoukar A, Ali J (2013) GHz frequency band soliton generation using integrated ring resonator for WiMAX optical communication. Opt Quantum Electron
19. Amiri IS, Alavi SE, Rahim FJ, Idrus SM (2014) Analytical treatment of the ring resonator passive systems and bandwidth characterization using directional coupling coefficients. J Comput Theor Nanosci
20. Amiri IS, Alavi SE, Bahadoran M, Afroozeh A, Idrus SM (2014) Nanometer bandwidth soliton generation and experimental transmission within nonlinear fiber optics using an add-drop filter system. J Comput Theor Nanosci
21. Amiri IS, Ghorbani S, Naraei P (2014) Chaotic carrier signal generation and quantum transmission along fiber optics communication using integrated ring resonators. Quantum Matter
22. Alavi SE, Amiri IS, Idrus SM, Supa'at ASM (2014) Generation and wired/wireless transmission of IEEE802.16m signal using solitons generated by microring resonator. Opt Quantum Electron
23. Afroozeh A, Amiri IS, Ali J, Yupapin PP (2012) Determination of fwhm for soliton trapping. Jurnal Teknologi 55:77–83
24. Ali J, Amiri IS, Afroozeh A, Kouhnavard M, Jalil M, Yupapin PP (2010) Simultaneous dark and bright soliton trapping using nonlinear MRR and NRR. ICAMN, International Conference, Prince Hotel, Kuala Lumpur, Malaysia
25. Ali J, Kouhnavard M, Amiri IS, Afroozeh A, Jalil MA, Naim I, Yupapin PP (2010) Localization of soliton pulse using nano-waveguide. ICAMN, International Conference, Prince Hotel, Kuala Lumpur, Malaysia
26. Amiri IS (2011) Optical soliton trapping for quantum key generation. The International Conference for Nano materials Synthesis and Characterization, International Atomic Energy Agency (IAEA), Malaysia
27. Amiri IS (2014) Multiplex and de-multiplex of generated multi optical soliton by MRRs using fiber optics transmission link. Quantum Matter
28. Amiri IS, Ali J (2014) Femtosecond optical quantum memory generation using optical bright soliton. J Comput Theor Nanosci 11(6):1480–1485
29. Alavi SE, Amiri IS, Idrus SM, Supa'at ASM, Ali J (2013) Chaotic signal generation and trapping using an optical transmission link. Life Sci J 10(9):186–192
30. Amiri IS, Nikmaram M, Shahidinejad A, Ali J (2013) Generation of potential wells used for quantum codes transmission via a TDMA network communication system. Secur Commun Netw 6(11):1301–1309
31. Teeka C, Songmuang S, Jomtarak R, Yupapin PP, Jalil M, Amiri IS, Ali J (2011) ASK-to-PSK Generation based on nonlinear microring resonators coupled to one MZI arm. AIP Conf Proc 1341(1):221–223
32. Afroozeh A, Amiri IS, Chaudhary K, Ali J, Yupapin PP (2014) Analysis of optical ring resonator. Advances in laser and optics research. Nova Science, New York
33. Ali J, Jalil M, Amiri IS, Afroozeh A, Kouhnavard M, Naim I, Yupapin PP (2010) Multi-wavelength narrow pulse generation using MRR. ICAMN, International Conference, Prince Hotel, Kuala Lumpur, Malaysia

34. Amiri IS, Ahsan R, Shahidinejad A, Ali J, Yupapin PP (2012) Characterisation of bifurcation and chaos in silicon microring resonator. IET Commun 6(16):2671–2675
35. Amiri IS, Raman K, Afroozeh A, Jalil MA, Nawi IN, Ali J, Yupapin PP (2011) Generation of DSA for security application. Procedia Eng 8:360–365
36. Amiri IS, Ali J (2014) Optical quantum generation and transmission of 57–61 GHz frequency band using an optical fiber optics. J Comput Theor Nanosci 11(10)
37. Amiri IS, Alavi SE, Ali J (2013) High capacity soliton transmission for indoor and outdoor communications using integrated ring resonators. Int J Commun Syst
38. Amiri IS, Soltanmohammadi S, Shahidinejad A, Ali J (2013) Optical quantum transmitter with finesse of 30 at 800-nm central wavelength using microring resonators. Opt Quantum Electron 45(10):1095–1105
39. Amiri IS, Alavi SE, Shahidinejad A, Nikoukar A, Anwar T, Supa'at ASM, Idrus SM, Yen NK (2014) Characterization of ultra-short soliton generation using MRRs. The 2014 Third ICT International Student Project Conference (ICT-ISPC2014). IEEE, Thailand
40. Ridha NJ, Mohamad FK, Amiri IS, Saktioto, Ali J, Yupapin PP (2010) Controlling center wavelength and free spectrum range by MRR radii. International Conference on Experimental Mechanics (ICEM), Kuala Lumpur, Malaysia
41. Afroozeh A, Amiri IS, Kouhnavard M, Jalil M, Ali J, Yupapin PP (2010) Optical dark and bright soliton generation and amplification. AIP Conf Proc 1341:259–263
42. Ali J, Raman K, Afroozeh A, Amiri IS, Jalil MA, Nawi IN, Yupapin PP (2010) Generation of DSA for security application. 2nd International Science, Social Science, Engineering Energy Conference (I-SEEC 2010), Nakhonphanom, Thailand
43. Amiri IS, Nikoukar A, Shahidinejad A, Ranjbar M, Ali J, Yupapin PP (2012) Generation of quantum photon information using extremely narrow optical tweezers for computer network communication. GSTF J Comput 2(1):140
44. Amiri IS (2011) FWHM Measurement of Localized Optical Soliton. The International Conference for Nano materials Synthesis and Characterization. International Atomic Energy Agency (IAEA), Malaysia
45. Amiri IS, Shahidinejad A (2014) Generating of 57–61 GHz frequency band using a panda ring resonator. Quantum Matter
46. Amiri IS, Ali J (2014) Simulation of the single ring resonator based on the Z-transform method theory. Quantum Matter 3(6):519–522
47. Nikoukar A, Amiri IS, Ali J (2013) Generation of nanometer optical tweezers used for optical communication networks. Int J Innovative Res Comput Commun Eng 1(1):77–85
48. Afroozeh A, Amiri IS, Zeinalinezhad A (2014) Micro ring resonators and applications. LAP LAMBERT Academic Publishing, Saarbrücken, Germany
49. Ali J, Amiri IS, Jalil M, Kouhnavard M, Afroozeh A, Naim I, Yupapin PP (2010) Narrow UV pulse generation using MRR and NRR system. ICAMN, International Conference. Prince Hotel, Kuala Lumpur, Malaysia
50. Ali J, Kouhnavard M, Afroozeh A, Amiri IS, Jalil MA, Yupapin PP (2010) Optical bistability in a FORR. ICEM, Legend Hotel, Kuala Lumpur, Malaysia
51. Amiri IS, Ali J (2014) Picosecond soliton pulse generation using a PANDA system for solar cells fabrication. J Comput Theor Nanosci 11(3):693–701
52. Amiri IS, Ranjbar M, Nikoukar A, Shahidinejad A, Ali J, Yupapin PP (2012) Multi optical soliton generated by PANDA ring resonator for secure network communication. Computer and Communication Engineering (ICCCE) Conference, IEEE Explore, Malaysia
53. Amiri IS, Nikoukar A, Shahidinejad A, Anwar T (2014) The proposal of high capacity GHz soliton carrier signals applied for wireless commutation. Rev Theor Sci 2(4):320–333
54. Amiri IS, Naraei P, Ali J (2014) Review and theory of optical soliton generation used to improve the security and high capacity of MRR and NRR passive systems. J Comput Theor Nanosci 11(9):1875–1886
55. Amiri IS, Alavi SE, Idrus SM (2015) RF signal generation and wireless transmission using PANDA and Add/drop systems. J Comput Theor Nanosci

56. Amiri IS, Alavi SE, Idrus SM, Nikoukar A, Ali J (2013) IEEE 802.15.3c WPAN standard using millimeter optical soliton pulse generated by a panda ring resonator. IEEE Photonics J 5(5):7901912
57. Amiri IS, Nikoukar A, Alavi SE (2014) Soliton and radio over fiber (RoF) applications. LAP LAMBERT Academic Publishing, Saarbrücken, Germany
58. Alavi SE, Amiri IS, Supa'at ASM, Idrus SM (2014) Indoor data transmission over ubiquitous infrastructure of powerline cables and LED lighting. J Comput Theor Nanosci
59. Suwanpayak N, Songmuang S, Jalil MA, Amiri IS, Naim I, Ali J, Yupapin PP (2010) Tunable and storage potential wells using microring resonator system for bio-cell trapping and delivery. AIP Conf Proc 1341:289–291
60. Zeinalinezhad A, Pourmand SE, Amiri IS, Afroozeh A (2014) Stop light generation using nano ring resonators for ROM. J Comput Theor Nanosci
61. Afroozeh A, Amiri IS, Jalil MA, Kouhnavard M, Ali J, Yupapin PP (2011) Multi soliton generation for enhance optical communication. Appl Mech Mater 83:136–140
62. Ali J, Amiri IS, Jalil A, Kouhnavard A, Mitatha B, Yupapin PP (2010) Quantum internet via a quantum processor. International Conference on Photonics (ICP 2010), Langkawi, Malaysia
63. Amiri IS, Afroozeh A, Ali J, Yupapin PP (2012) Generation of quantum codes using up and down link optical solition. Jurnal Teknologi 55:97–106
64. Amiri IS, Ali J (2013) Nano optical tweezers generation used for heat surgery of a human tissue cancer cells using add/drop interferometer system. Quantum Matter 2(6):489–493
65. Amiri IS, Khanmirzaei MH, Kouhnavard M, Yupapin PP, Ali J (2012) Quantum entanglement using multi dark soliton correlation for multivariable quantum router. In: Moran AM (ed) Quantum entanglement. Nova Science Publisher, New York, pp 111–122
66. Amiri IS, Nikoukar A, Ali J, Yupapin PP (2012) Ultra-short of pico and femtosecond soliton laser pulse using microring resonator for cancer cells treatment. Quantum Matter 1(2):159–165
67. Amiri IS, Alavi SE, Idrus SM, Afroozeh A, Ali J (2014) Soliton generation by ring resonator for optical communication application. Nova Science Publishers, USA
68. Amiri IS, Alavi SE, Idrus SM (2014) Soliton coding for secured optical communication link. Springer, USA
69. Shahidinejad A, Nikoukar A, Amiri IS, Ranjbar M, Shojaei A, Ali J, Yupapin PP (2012) Network system engineering by controlling the chaotic signals using silicon micro ring resonator. Computer and Communication Engineering (ICCCE) Conference, IEEE Explore, Malaysia
70. Ali J, Afroozeh A, Amiri IS, Jalil M, Yupapin PP (2010) Wide and narrow signal generation using chaotic wave. Nanotech Malaysia, International Conference on Enabling Science & Technology. Kuala Lumpur, Malaysia
71. Shahidinejad A, Amiri IS, Anwar T (2014) Enhancement of indoor wavelength division multiplexing-based optical wireless communication using microring resonator. Rev Theor Sci 2(3):201–210
72. Amiri IS, Alavi SE, Idrus SM (2015) Introduction of fiber waveguide and soliton signals used to enhance the communication security. Soliton coding for secured optical communication link. Springer, USA, pp 1–16
73. Amiri IS, Ali J (2012) Generation of nano optical tweezers using an add/drop interferometer system. 2nd Postgraduate Student Conference (PGSC), Singapore
74. Amiri IS, Ali J, Yupapin PP (2012) Enhancement of FSR and finesse using add/drop filter and PANDA ring resonator systems. Int J Mod Phys B 26(04):1250034
75. Amiri IS, Afroozeh A (2014) Spatial and temporal soliton pulse generation by transmission of chaotic signals using fiber optic link advances in laser and optics research, vol 11. Nova Science Publisher, New York
76. Amiri IS, Nikoukar A, Shahidinejad A, Anwar T, Ali J (2014) Quantum transmission of optical tweezers via fiber optic using half-panda system. Life Sci J 10(12):391–400
77. Amiri IS, Ebrahimi M, Yazdavar AH, Gorbani S, Alavi SE, Idrus SM, Ali J (2014) Transmission of data with orthogonal frequency division multiplexing technique for communication networks using GHz frequency band soliton carrier. IET Commun 8(8):1364–1373

78. Amiri IS, Alavi SE, Idrus SM, Supa'at ASM, Ali J, Yupapin PP (2014) W-band OFDM transmission for radio-over-fiber link using solitonic millimeter wave generated by MRR. IEEE J Quantum Electron 50(8):622–628
79. Amiri IS, Rahim FJ, Arif AS, Ghorbani S, Naraei P, Forsyth D, Ali J (2014) Single soliton bandwidth generation and manipulation by microring resonator. Life Sci J 10(12):904–910
80. Amiri IS, Alavi SE, Idrus SM, Kouhnavard M (2014) Microring resonator for secured optical communication. Amazon, USA
81. Nikoukar A, Amiri IS, Shahidinejad A, Shojaei A, Ali J, Yupapin PP (2012) MRR quantum dense coding for optical wireless communication system using decimal convertor. Computer and Communication Engineering (ICCCE) Conference, IEEE Explore, Malaysia
82. Naraei P, Amiri IS, Saberi I (2014) Optimizing IEEE 802.11i resource and security essentials for mobile and stationary devices. Elsevier
83. Alavi SE, Amiri IS, Idrus SM, Supa'at ASM (2014) Optical amplification of tweezers and bright soliton using an interferometer ring resonator system. J Comput Theor Nanosci
84. Alavi SE, Amiri IS, Idrus SM, Ali J A (2013) Optical wired/wireless communication using soliton optical tweezers. Life Sci J 10(12s):179–187

Chapter 4
Soliton Generation Based Optical Communication

Abstract In this chapter, required optical soliton pulses are generated using integrated ring resonators. These pulses can be used in optical communication, where the quality of the optical transmission can be improved significantly using soliton pulses. The results of chaotic signals, dark and bright soliton generation and conversion, single and multiple soliton, dark soliton array and entangle photons, multisoliton using a PANDA ring resonator system, and the finesse improvement are presented using the Z-transform and iterative methods.

Keywords Optical Soliton · Chaotic · Bright soliton · Dark soliton · Dark soliton array · Entangle photons · Finesse

4.1 Introduction

In this chapter, simulated results from chaotic signals, dark and bright soliton generation and conversion, single and multiple soliton, dark soliton array and entangle photon generation, the multisoliton generation using a PANDA ring resonator system and the way of finesse improvement are presented. The numerical software used is the MATLAB.

4.2 Disordered Soliton Pulses (Chaotics)

A bright soliton pulse with 20 ns pulse width, peak power of 500 mW is input into the series of three ring resonators shown in Fig. 4.1. The suitable ring parameters are used, for instance, ring radii $R_1 = R_2 = R_3 = 10$ μm. The selected parameters of the system are fixed to $\lambda_0 = 1.5$ μm, $n_0 = 3.34$ (InGaAsP/InP), $A_{eff} = 0.50$, 0.25 and 0.12 μm^2 for different radii of microring resonators respectively, $\alpha = 0.5$ dB mm^{-1}, $\gamma = 0.1$. The coupling coefficient (kappa, κ) of the microring resonator ranged from 0.3 to 0.7.

© The Author(s) 2015
I. Sadegh Amiri and A. Afroozeh, *Ring Resonator Systems to Perform Optical Communication Enhancement Using Soliton*, SpringerBriefs in Applied Sciences and Technology, DOI 10.1007/978-981-287-197-8_4

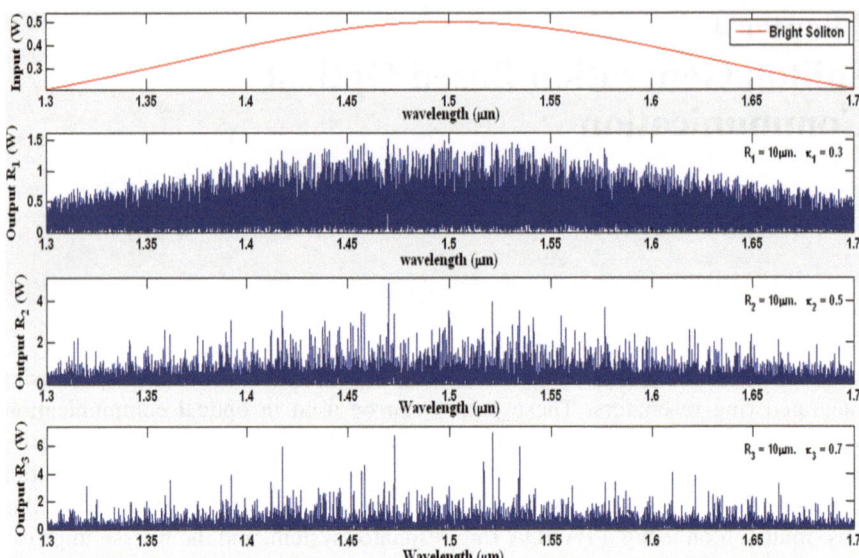

Fig. 4.1 Simulation results of spatial chaotic signals generation within a series of microring resonators with bright soliton input

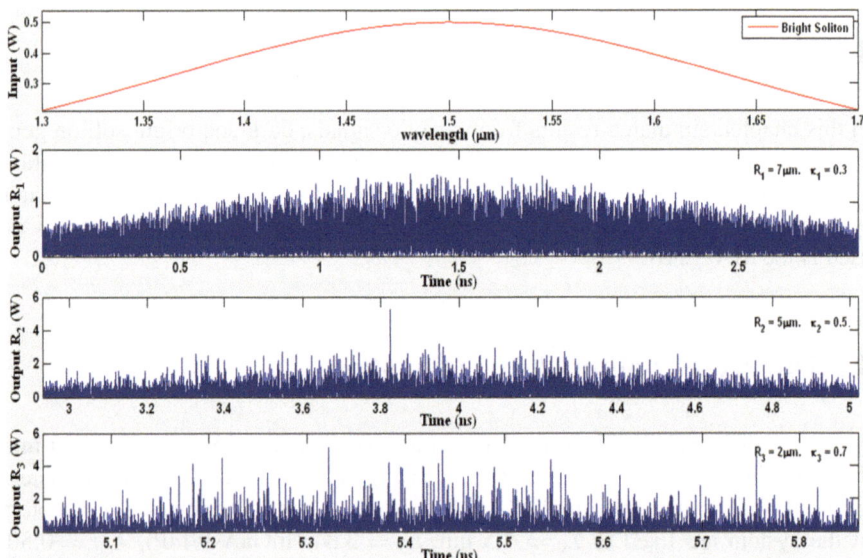

Fig. 4.2 Simulation results of temporal chaotic signals generation within a series of microring resonators with bright soliton input

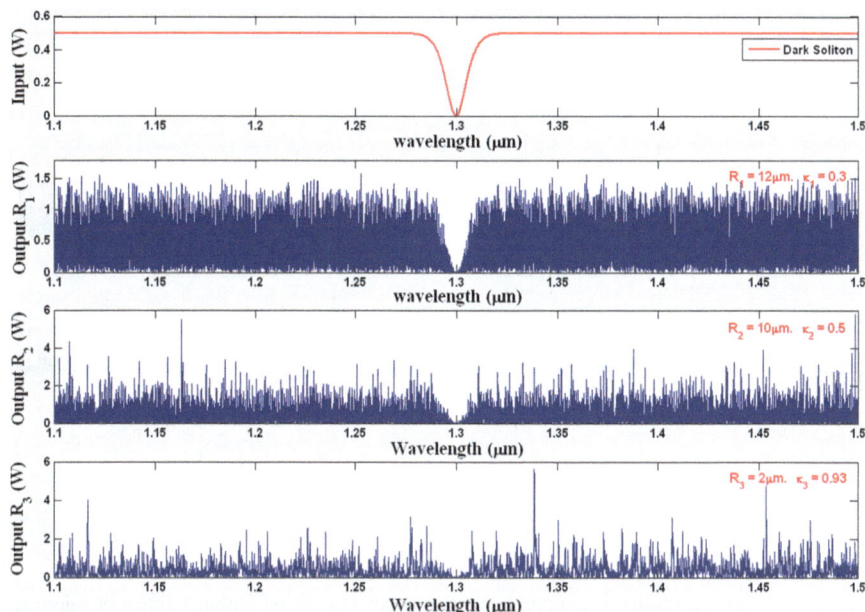

Fig. 4.3 Simulation results of spatial chaotic signals generation within a series of microring resonators with dark soliton input

The time depends of the output chaotic signals which is shown in Fig. 4.2, where the radii of the rings have been selected to $R_1 = 7\,\mu m$, $R_2 = 5\,\mu m$, and $R_3 = 2\,\mu m$.

The input pulse of ultra-short dark soliton with a power of 500 mW, central wavelength of $\lambda_0 = 1.5\,\mu m$ and bandwidth of 12 nm can be inserted into the system, where the signals of chaos can be generated shown in Fig. 4.3. The radii of the ring resonators are selected to $R_1 = 12\,\mu m$, $R_2 = 10\,\mu m$ and $R_3 = 2\,\mu m$, where $\kappa_1 = 0.3$, $\kappa_2 = 0.5$ and $\kappa_3 = 0.93$.

Temporal profile of the input dark soliton pulse can be seen from Fig. 4.4, where the input of the pulse is 350 mW with central wavelength of 1.3 μm. The ring radii and coupling coefficients of the rings are selected to $R_1 = 30\,\mu m$, $R_2 = 12\,\mu m$ and $R_3 = 5\,\mu m$, where $\kappa_1 = 0.7$, $\kappa_2 = 0.9$ and $\kappa_3 = 0.93$.

4.3 Dark Form of Soliton Generation

Optical field of the Gaussian pulse can be inserted into the input port of the multi-stage MRR's system. Considering the proposed system, the radii of the rings have been selected as $R_1 = 15\,\mu m$, $R_2 = 10\,\mu m$, $R_3 = 5\,\mu m$, and $\kappa_1 = 0.5$, $\kappa_2 = 0.7$, $\kappa_3 = 0.7$, where the add/drop filter has a radius of $R_d = 50\,\mu m$ and coupling coefficients of $\kappa_4 = \kappa_5 = 0.3$. Some parameters of the system are fixed such as $n_0 = 3.34$ (InGaAsP/InP), $A_{eff} = 0.50$, 0.250 and 0.10 μm^2 for the microrings, $\alpha = 0.5$ dB mm^{-1}, $\gamma = 0.1$. The nonlinear refractive index of the microring is $n_2 = 2.2 \times 10^{-17}\,m^2/W$ [1–7].

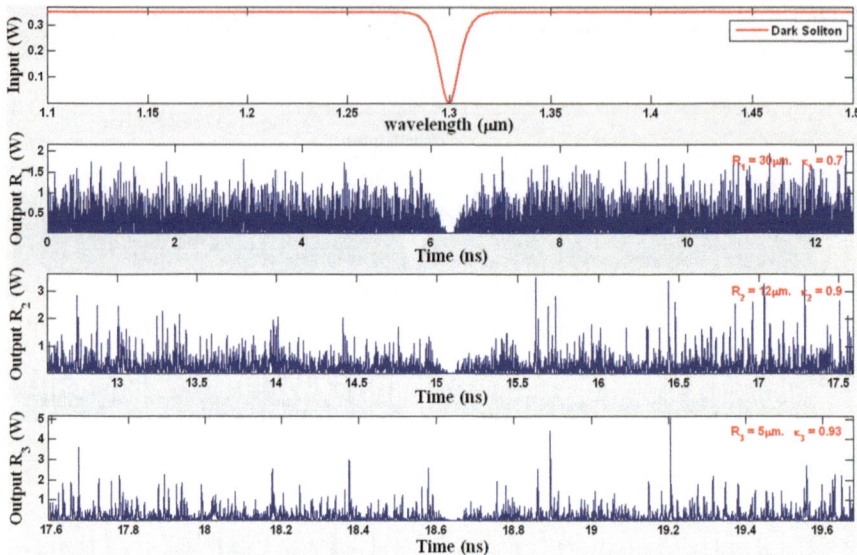

Fig. 4.4 Simulation results of temporal chaotic signals generation within a series of microring resonators with dark soliton input

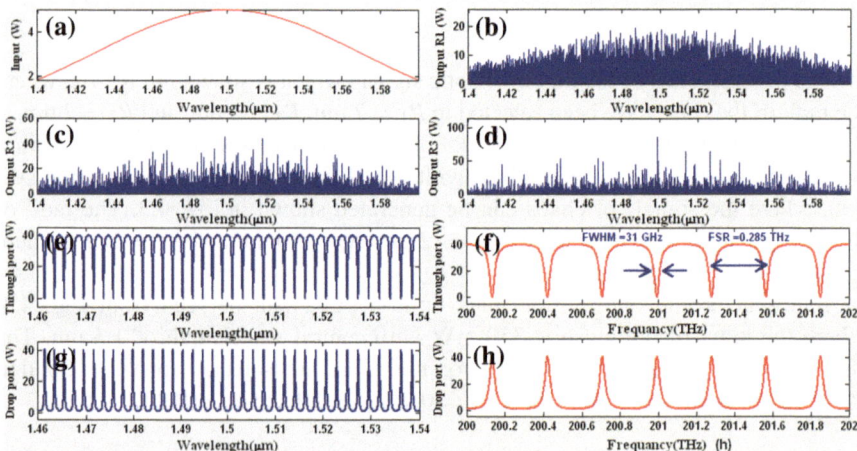

Fig. 4.5 Simulation results of spatial dark and bright soliton generation, where **a** input Gaussian beam, **b** chaotic signals from first ring, **c** chaotic signals from second ring, **d** chaotic signals from third ring, **e** spatial dark soliton at the through port, **f** THz signals at the through port with FWHM and FSR of 31 GHz and 0.285 THz respectively, **g** bright soliton generation at the drop port with FWHM and FSR of 31 GHz and 0.285 THz respectively, **h** THz signals generation at the drop port

The input Gaussian laser pulse with power of 5 W is introduced into the MRR's system shown in Fig. 4.5a. The output powers from three ring resonators are shown in Fig. 4.5b–d, where Fig. 4.5e, f show the output power from the throughput port in terms of wavelength and frequency respectively. The FWHM and FSR of the

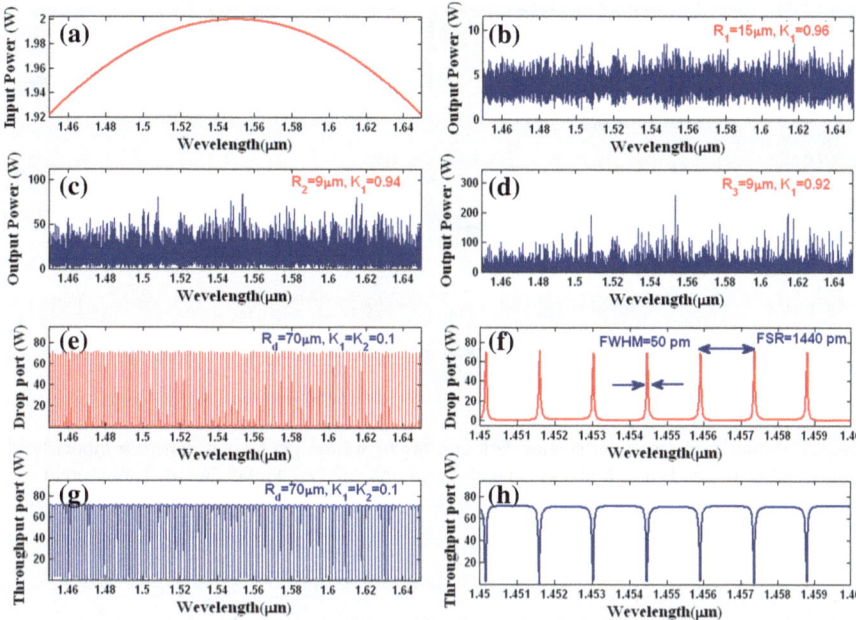

Fig. 4.6 Results of the multisoliton pulse generation, **a** input bright soliton, **b–d** large bandwidth signals, **e–f** bright soliton with FSR of 1,440 pm, and FWHM of 50 pm, **g–h** dark soliton with FSR of 1,440 pm, and FWHM of 50 pm

spatial soliton pulses are 31 GHz and 0.285 THz respectively. Figure 4.5g, h show the output power from the drop port of the system with the same FWHM and FSR.

In Fig. 4.6, the input bright soliton pulse has 50 ns pulse width and peak power of 2 W. The ring radii are $R_1 = 15$ μm, $R_2 = 9$ μm, $R_3 = 9$ and $R_d = 70$ μm. The fixed parameters are selected to $\lambda_0 = 1.55$ μm, $n_0 = 3.34$ (InGaAsP/InP), $A_{eff} = 0.25$ μm², $\alpha = 0.5$ dB mm^{-1}, $\gamma = 0.1$. The coupling coefficients range from 0.1 to 0.96, where the nonlinear refractive index is $n_2 = 2.2 \times 10^{-17}$ m²/W and the wave guided loss used is 0.5 dB mm^{-1}. Optical signals are sliced into smaller signals broadening over the band as shown in Fig. 4.6b–d. Therefore, large bandwidth signal is formed within the first ring device, where compress bandwidth with smaller group velocity is attained inside the ring R_2 and R_3, such as filtering signals. Localized soliton pulses are formed within the add/drop filter system, where resonant condition is performed, given in Fig. 4.6e–h. However, there are two types of dark and bright soliton pulses. Here the multi dark soliton pulses with FSR and FHWM of 1,440 and 50 pm are simulated.

To control the output signals from the through and drop ports of the add/drop filter system, an additional input signal can be added to the add port shown in Fig. 4.7. Optical fields of bright soliton and Gaussian pulse with input powers of 3 and 2 W insert into the input and add ports of the system respectively. The radii of the rings are selected to $R_1 = 15$ μm, $R_2 = 9$ μm, $R_3 = 5$ μm, and $\kappa_1 = 0.5$, $\kappa_2 = 0.6$, $\kappa_3 = 0.7$, where the add/drop filter has a radius of $R_d = 700$ μm and

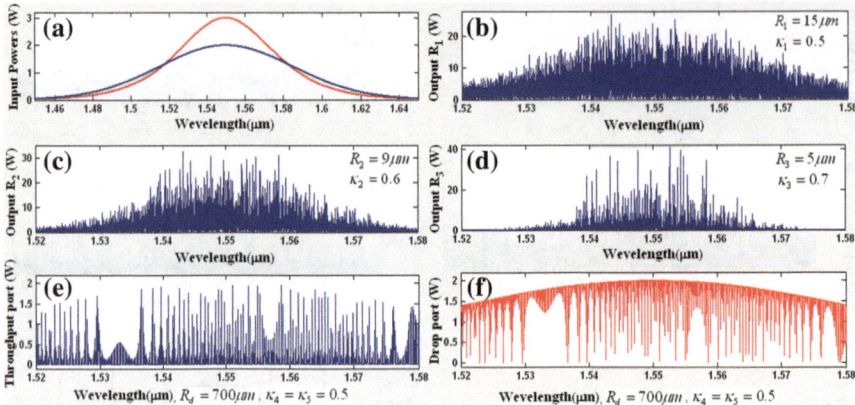

Fig. 4.7 Simulation results of spatial dark and bright soliton generation, where **a** inputs bright soliton and Gaussian beam, **b** chaotic signals from first ring, **c** chaotic signals from second ring, **d** chaotic signals from third ring, **e** bright soliton generation at the through port, **f** dark soliton generation at the drop port

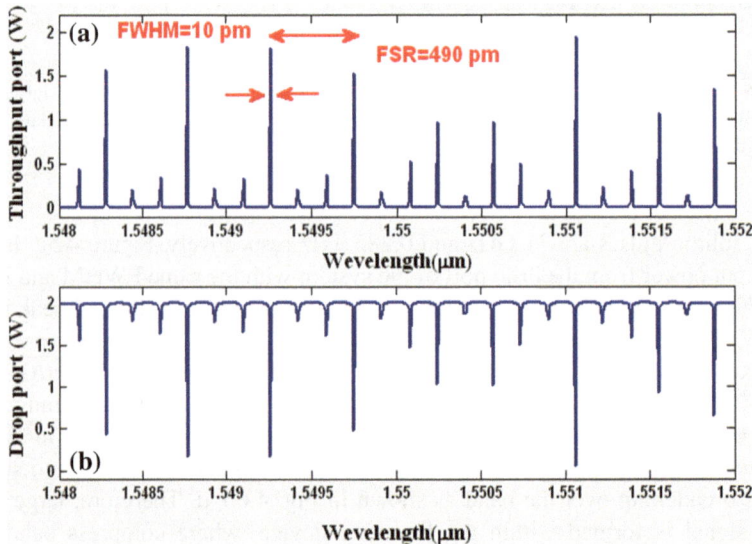

Fig. 4.8 Simulation results of spatial dark and bright soliton generation, where **a** bright soliton generation at the through port with FWHM and FSR of 10 and 490 pm respectively, **b** dark soliton generation at the drop port with FWHM and FSR of 10 and 490 pm respectively

coupling coefficients of $\kappa_4 = \kappa_5 = 0.5$. Here multi bright and dark soliton with FWHM and FSR of 10 and 160 pm could be generated respectively.

The nonlinear refractive index of the microring is $n_2 = 3.4 \times 10^{-17}$ m^2/W. Here dark and bright soliton with FWHM and FSR of 10 and 280 pm are simulated shown in Fig. 4.8.

4.4 Ultra-Short Solitons

In operation, the large bandwidth within the microring device can be generated by using a soliton pulse input into the nonlinear MRR system [8–12], where the required signals can perform the generation of ultra-short soliton pulses. The nonlinear refractive index is selected to $n_2 = 2.2 \times 10^{-17}$ m^2/W. As it is shown in Fig. 4.9, the input signal is inserted to the rings system, where it will be chopped (sliced) into smaller signals spreading over the spectrum, which shows that the large bandwidth is formed within the first ring device [13–17]. The compress bandwidth with smaller group velocity is obtained within the ring R_2. The amplifier gain is obtained within the R_3 microring device. The temporal soliton pulse can be formed by using constant gain condition, where a small group velocity is seen. Attenuation of the optical power within a microring device is required in order to keep the constant output gain. In this case, the time belongs to each ring resonator [18–25].

Similarly, trapping of spatial soliton at the 1,557 nm of wavelength is obtained as shown in Fig. 4.10, where the FWHM of the pulse is 17.5 pm.

Results of femtosecond pulse generation can be seen from Fig. 4.11, where the radius of the rings has been selected to $R_1 = 17\,\mu$m, $R_2 = 13\,\mu$m, $R_3 = 7\,\mu$m, $R_4 = 2\,\mu$m. The center wavelength of the input bright soliton is $\lambda = 0.6\,\mu$m.

Spatial soliton pulse with FWHM of 0.34 nm can be generated at $\lambda = 555$ nm shown in Fig. 4.12d where the steps of filtering process are shown in Fig. 4.12a–c.

Figure 4.13 shows the results of temporal and spatial optical soliton pulses localized within a microring device and add/drop filter system with 20,000 round-trip, where an optical ultra-short temporal soliton of FWHM = 83 fs is generated. Here, the ring radii are $R_1 = 10\,\mu$m, $R_2 = 5\,\mu$m, $R_3 = 4\,\mu$m, $R_4 = 4\,\mu$m and

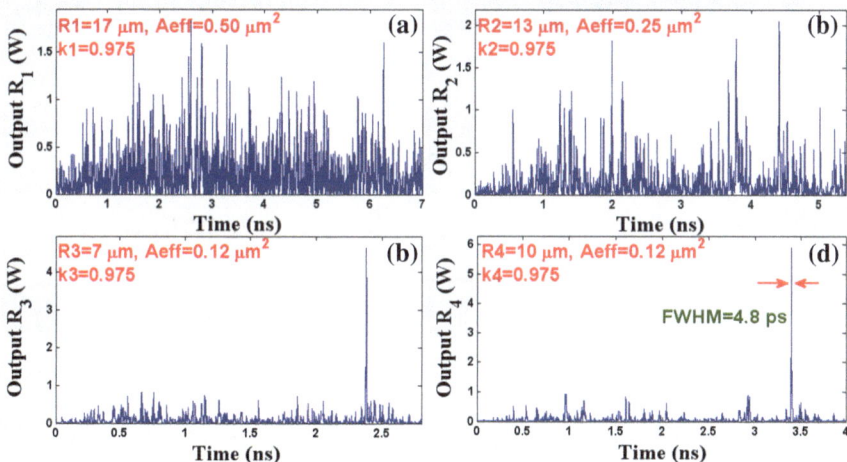

Fig. 4.9 Results obtained when temporal soliton is localized within a microring device with 20,000 roundtrips, where **a** chaotic signals from R_1, **b** chaotic signals from R_2, **c** localized temporal soliton, **d** localized temporal soliton with FWHM of 4.8 ps

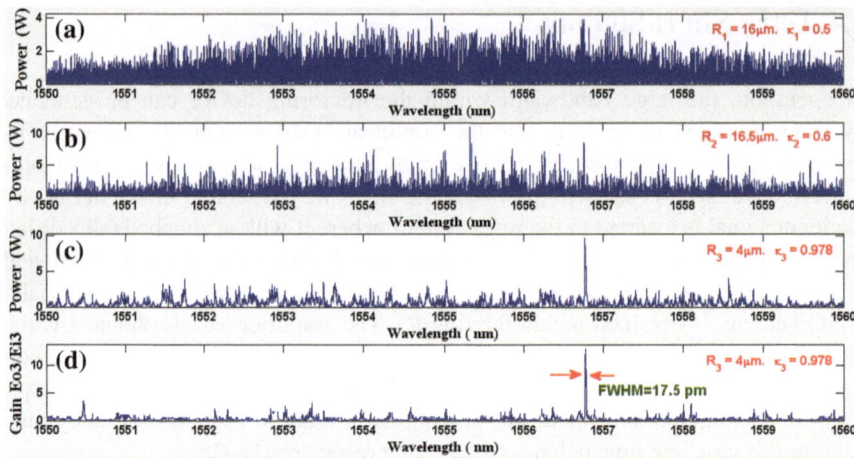

Fig. 4.10 Results of trapping spatial soliton within a microring device with 20,000 roundtrips, where **a** chaotic signals from R_1, **b** chaotic signals from R_2, **c** localized spatial soliton, **d** localized spatial soliton with FWHM of 17.5 pm

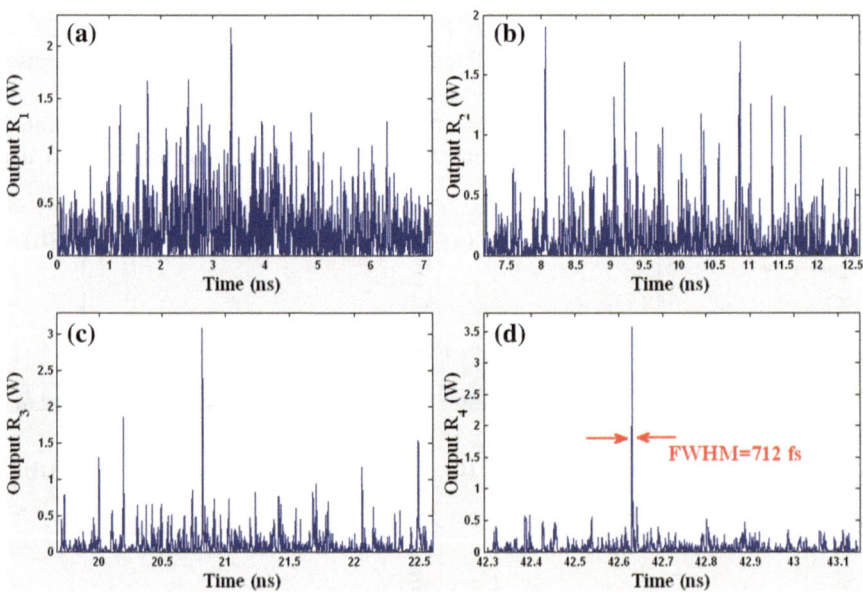

Fig. 4.11 Results obtained when temporal soliton is localized within a microring device with 20,000 roundtrips, where **a** chaotic signals from R_1, **b** chaotic signals from R_2, **c** localized temporal soliton, **d** localized temporal soliton with FWHM of 712 fs

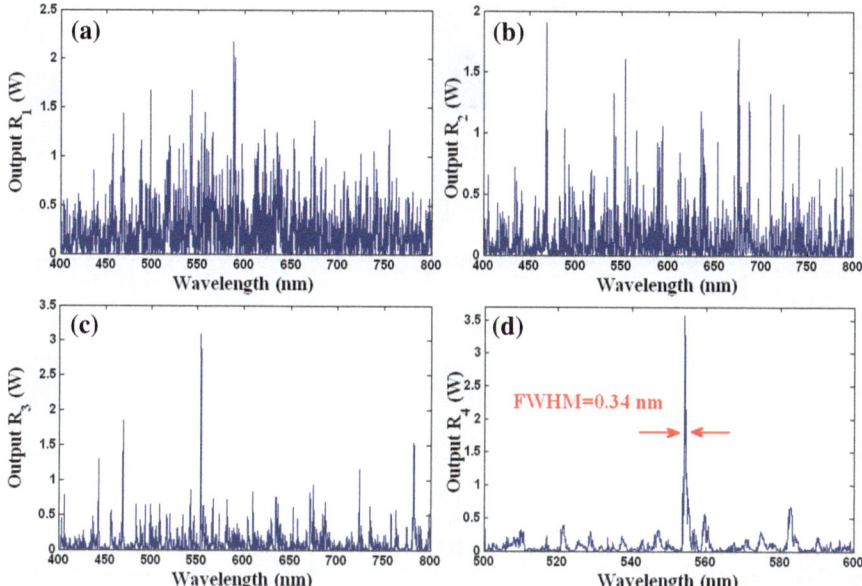

Fig. 4.12 Results of localized spatial soliton within a microring device with 20,000 round-trip, where **a** chaotic signal from R_1, **b** chaotic signals from R_2, **c** localized spatial soliton, **d** localized spatial soliton with FWHM of 0.34 nm

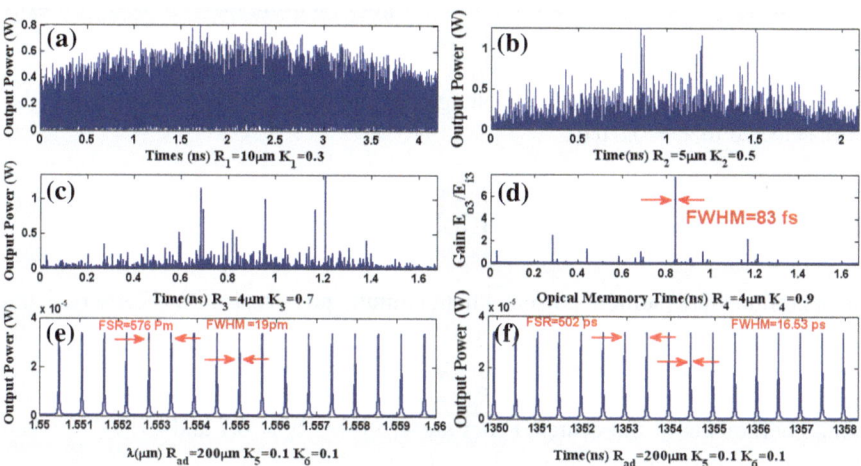

Fig. 4.13 Results of temporal and spatial soliton generation, where **a** chaotic signals from R_1, **b** chaotic signals from R_2, **c** filtering signals, **d** localized temporal soliton with FWHM of 83 fs, **e** spatial soliton with FSR = 576 pm and FWHM = 19 pm, **f** temporal soliton with FSR = 502 ps and FWHM = 16.53 ps

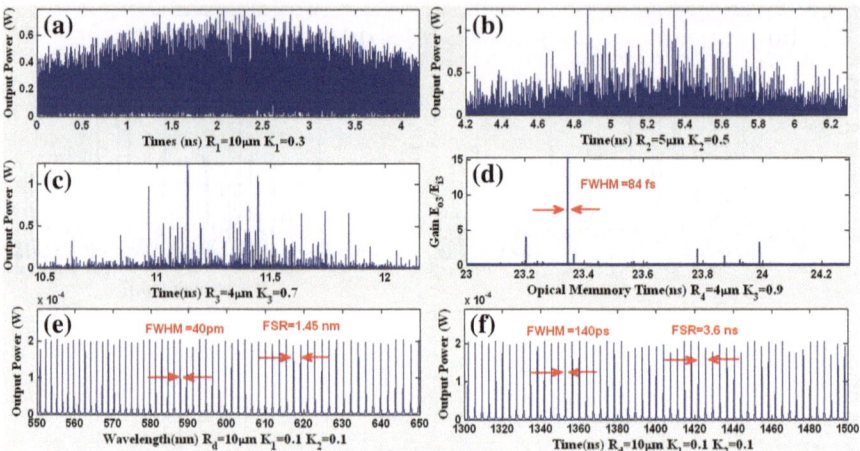

Fig. 4.14 Results of temporal and spatial soliton generation, where **a** chaotic signals from R_1, **b** chaotic signals from R_2, **c** filtering signals, **d** localized temporal soliton with FWHM of 84 fs, **e** spatial soliton with FSR = 1.45 nm and FWHM = 40 pm, **f** temporal soliton with FSR = 3.6 ns and FWHM = 140 ps

$R_{ad} = 200 \, \mu m$ with coupling coefficient of $\kappa_1 = 0.3$, $\kappa_2 = 0.5$, $\kappa_3 = 0.7$, $\kappa_4 = 0.9$, $\kappa_5 = 0.1$ and $\kappa_6 = 0.1$.

Results of localized ultra-short temporal optical soliton with FWHM of 84 fs is shown in Fig. 4.14, where the multi temporal soliton with FWHM = 140 ps and FSR = 3.6 ns are obtained. The multi spatial solitons have FWHM = 40 pm and FSR = 1.45 nm, where the ring radius of the add/drop filter system is $R_{ad} = 10 \, \mu m$ and the central wavelength of the input bright soliton power has been selected to $\lambda = 0.6 \, \mu m$.

4.5 Soliton Array Using Multiple Wavelength

In operation dark soliton pulse with maximum power of 1 W is inserted into the input port of the interferometer PANDA ring resonator system, where the Gaussian beam power of 600 mW is input into the add port. The suitable ring parameters are ring radii, where $R_{ad} = 100 \, \mu m$ and $R_L = 800$ nm. The coupling coefficients of the centered ring are given by $\kappa_1 = 0.7$ and $\kappa_2 = 0.2$, where the ring resonator at the left side has coupling coefficient of $\kappa_3 = 0.35$. In order to make the system associate with the practical device, the selected parameters of the system are fixed to $\lambda_0 = 0.6 \, \mu m$, $n_0 = 3.34$ (InGaAsP/InP). The effective core areas range from $A_{eff} = 0.50$ to $0.10 \, \mu m^2$. The nonlinear refractive index is $n_2 = 1.3 \times 10^{-13} \, m^2/W$ [26–32].

After the Gaussian pulse is added into the system via add port, therefore the dark-Gaussian soliton collision is seen wherein extremely narrow dark soliton can be generated shown in Fig. 4.15. Figure 4.15a shows the inserted dark soliton and Gaussian

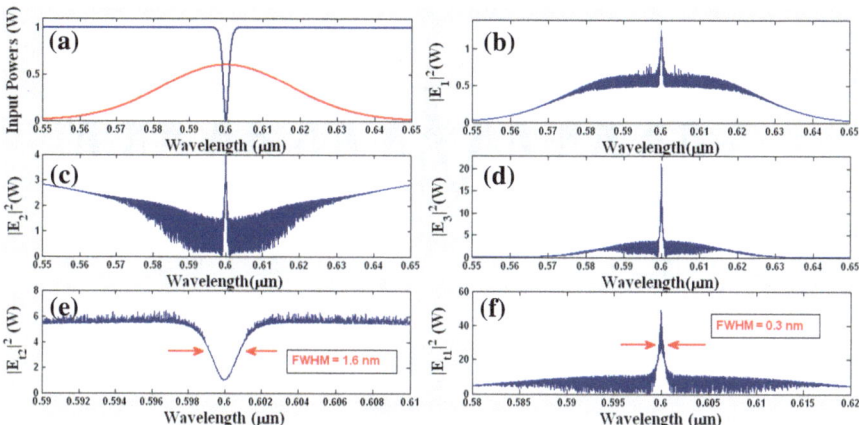

Fig. 4.15 Results of the dark soliton generation **a** input dark soliton and Gaussian pulse, **b–d** interior signals, **e, f** through and drop port output signals with FWHM of 0.3 and 1.6 nm respectively, where $K_1 = 0.7$, $K_2 = 0.2$, $K_3 = 0.35$

pulse at the input and add ports of the system with center wavelength of $\lambda_0 = 0.6\,\mu\text{m}$. Figure 4.15b–d shows the interior generated pulses with different amplitudes. The output signals from the throughput and drop ports of the system can be seen in Fig. 4.15e, f respectively. Here single bright and dark soliton with FWHM of 0.3 and 1.6 nm are simulated for the throughput and drop ports of the system respectively.

In order to achieve high capacity of a transmission, the optical dark soliton with different center wavelengths can be combined using the suitable multiplexer device [33–39]. Here, a series of MRR systems can be integrated in one single system, incorporating with multiplexer device. Highly potential well signals can be obtained from the output port of the multiplexer device shown in Fig. 4.16. Therefore, signals with center wavelengths of $\lambda_1 = 1.53\,\mu\text{m}$, $\lambda_2 = 1.535\,\mu\text{m}$, $\lambda_3 = 1.54\,\mu\text{m}$, $\lambda_4 = 1.545\,\mu\text{m}$, $\lambda_5 = 1.55\,\mu\text{m}$, $\lambda_6 = 1.555\,\mu\text{m}$, $\lambda_7 = 1.56\,\mu\text{m}$, $\lambda_8 = 1.565\,\mu\text{m}$ and $\lambda_9 = 1.57\,\mu\text{m}$ are combined, where pulses with FWHM and FSR of 0.8 and 5 nm could be obtained respectively.

In order to generate quantum binary and logic codes of "0" and "1", the multiplexed signals from the multiplexer system transmit into a beam splitter (PBS). In operation, the binary codes are generated within the proposed system after traveling of the signals through the PBS, whereas the polarization phase shift of the two components is 90°. This means that random polarization states of the two components can be used to form the binary code patterns and the binary code signals, which can be observed by using the photo detectors D_1 and D_2. The output bright and dark soliton signals from the beam splitter are shown in Fig. 4.17. The FWHM and FSR of the bright and dark soliton signals are 0.54 and 4.71 nm suitable for the digital code generation.

Therefore, the binary codes are generated in the form of dark and bright soliton using beam splitter [40–45]. The patterns of dark and bright solitons are '101', is

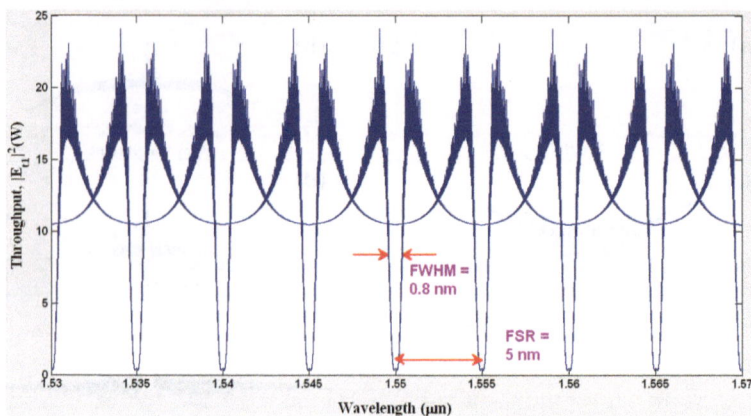

Fig. 4.16 Soliton array generation with FWHM and FSR of 0.8 and 5 nm respectively, using multiplexer system

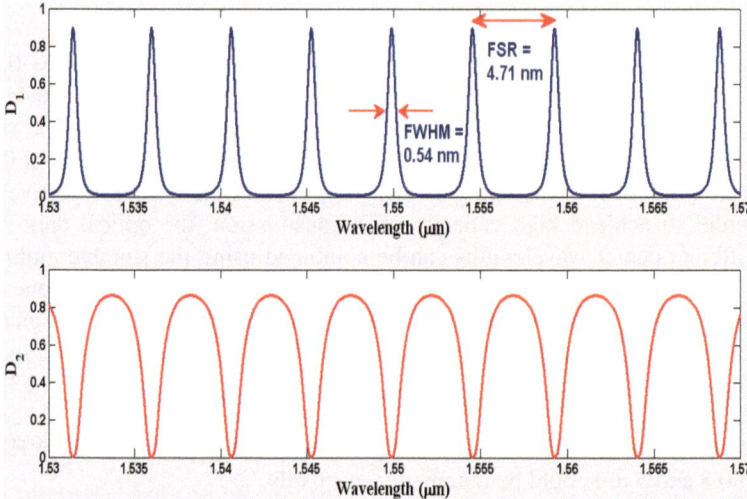

Fig. 4.17 Generation of bright and dark soliton signals (binary codes), using PBS, where D_1 and D_2 are signals detected by photo detector D_1 and D_2

'010'. The referencing binary code patterns and logic states are set as shown in Fig. 4.18a, b respectively.

Here, the logic codes detected by D_1 are '10101010101010101' patterns. The logic codes detected by D_2 are '01010101010101010' patterns. Different orders of the logic codes can be made to generate different signal information and propagated into the optical network communication.

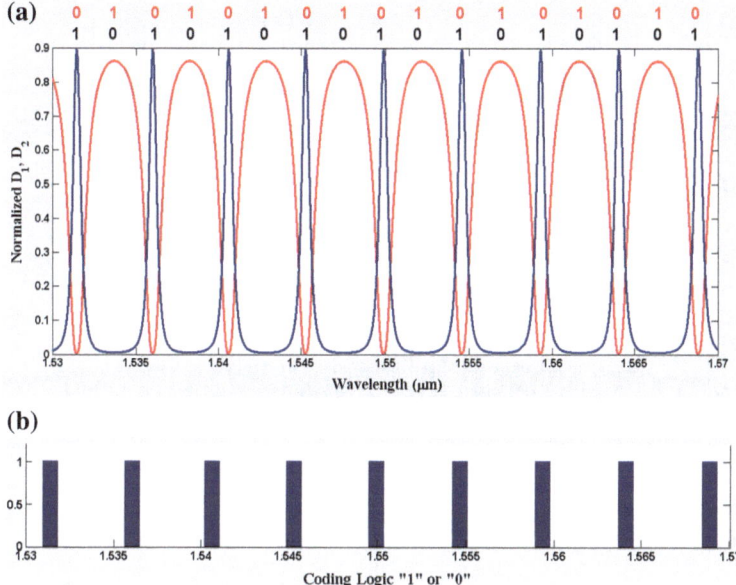

Fig. 4.18 Generation of binary and logic codes, using PBS, where **a** D_1 and D_2 are signals detected by photo detectors D_1 and D_2, **b** logic states of the binary codes

4.6 Multiple Soliton Generated by Ring Resonators

The system of nonlinear microring resonators can be used to generate multiple soliton channels. We propose a PANDA ring resonator system connected to an add/drop filter system with appropriate parameters. Gaussian beams with central wavelengths of 1.55 μm and powers of 600 mW are introduced into the add and input ports of the PANDA ring resonator. Simulated result of the nonlinear equations for the input power, propagating inside the fiber system has been shown in Fig. 4.19. The nonlinearity of the fiber system is on the Kerr effect type, where the linear and nonlinear refractive indices of the system are $n_0 = 3.34$ and $n_2 = 3.2 \times 10^{-17}$ respectively [46–51].

In Fig. 4.19, the coupling coefficients of the PANDA ring resonator are given as $\kappa_0 = 0.2$, $\kappa_1 = 0.35$, $\kappa_2 = 0.1$ and $\kappa_3 = 0.95$, respectively and $\gamma = \gamma_1 = \gamma_2 = 0.1$. The radius of the centered ring of PANDA system has been selected to $R_{PANDA} = 300$ nm where the radii of the right and left rings are $R_r = 180$ nm and $R_L = 200$ nm respectively. The output soliton signals are amplified and tuned using the add port of the system [52–59]. Figure 4.19a, b shows the powers in the form of chaotic signals before entering the right ring of the PANDA system and amplification of signals during propagation of light inside right ring respectively [60–67], where Fig. 4.19c, d shows the powers before entering the left ring and amplification of signals within the right ring respectively. We found that the

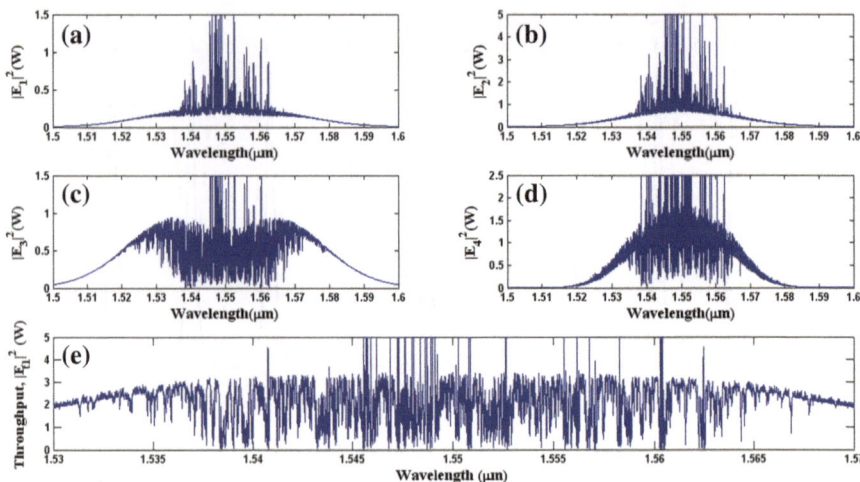

Fig. 4.19 Multisoliton signal generation using PANDA ring resonator system where **a–d** are powers inside the PANDA system and **e** is the output power from the throughput

signals are stable and seen within the system where the chaotic signals are generated at the through port shown in Fig. 4.19e.

Chaotic signals can be used in secured optical communication in which the information is input into the signals [68–74]. In order to retrieve the information from the chaotic signals, an add/drop filter system is used [75–80]. This system filters the chaotic signals and generates multisoliton which is used to improve the capacity of the system applicable for long distance communication. In order to generate multisoliton, the chaotic signals from the PANDA ring resonator are input into the add/drop filter system. Therefore the proposed system is suitable to enhance both the security and the capacity of the optical signals. Figure 4.20a, b shows the generation of multisoliton in the form of dark soliton and expansion of the through port signals respectively, where Fig. 4.20c, d represent multisoliton in the form of bright solitons and expansion of the drop port signals respectively. The coupling coefficients of the add/drop filter system are given as $\kappa_4 = 0.9$, $\kappa_5 = 0.5$, where the radius of the ring is $R_{ad} = 100\,\mu$m.

Apart from communication applications, the idea of personnel wavelength (network) is practical for the large demand user due to un-limited wavelength discrepancy, whereas the specific soliton band can be generated using the proposed system. The potential of soliton bands can be generated and used for many applications such as multi color holography, medical tools, security imaging and transparent holography and detection, respectively [81–86].

The system of dark soliton generation can be used to generate a high capacity of multi dark and bright soliton where pulses with FWHM and FSR of 50 pm and 1.44 nm can be simulated respectively shown in Fig. 4.21. The advantage of using this system is that pulses with larger FSR and smaller FWHM can be generated which is used to improve the sensitivity of the microring systems. Figure 4.21

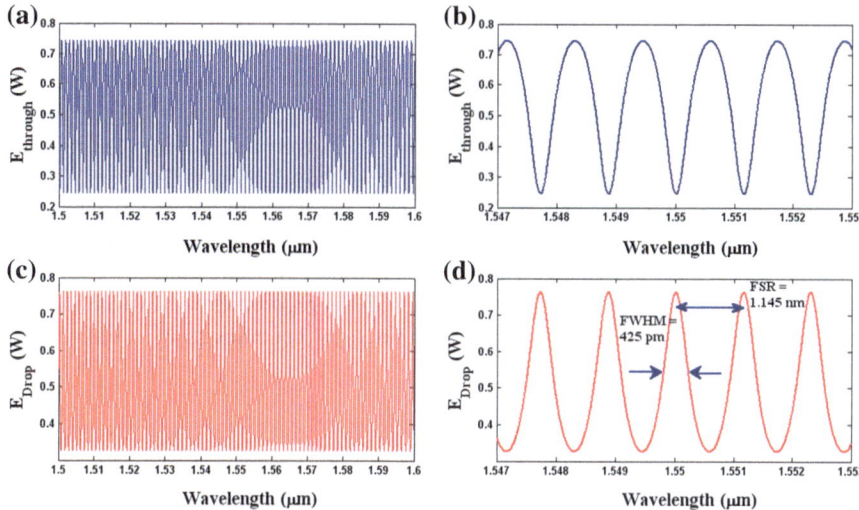

Fig. 4.20 Output multisoliton signal generation using an add/drop filter system, where **a** dark soliton at through port, **b** expansion of the multi dark soliton, **c** bright soliton at drop port, and **d** expansion of multi bright soliton with FWHM and FSR of 425 and 1.145 nm respectively

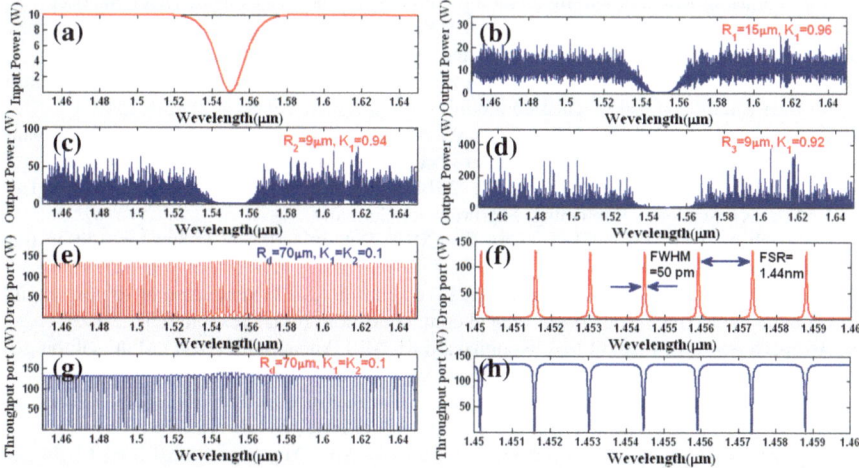

Fig. 4.21 Results of the multisoliton pulse generation, **a** input dark soliton, **b–d** large bandwidth signals, **e, f** bright soliton, **g, h** dark soliton with FSR of 1.44 nm, and FWHM of 50 pm

shows the generation of optical multisoliton signals, where the input soliton pulse of dark type with power of 10 W propagates inside the microring system. Figure 4.21b–d shows the filtering process within R_1, R_2 and R_3 with ring radii of 15, 9 and 9 μm respectively, where the amplification of output signals occurred. Using the dark soliton input power; the security of the optical transmission can be performed along the fiber due to very low power of the central wavelength, which secures the signals to be undetected by any types of conventional laser detectors. Therefore, by using suitable add/drop filter system, high power bright and dark soliton can be generated shown in Fig. 4.21e–h. Amplification of optical soliton is performed used to long transmission link. The power distribution of the output pulses can be executed via the add/drop filter with radius of R_d.

Therefore, the proposed system is suitable for the multisoliton pulse generation, which is available for high performance network [87–94]. Since, an optical soliton communication has been realized as a good candidate for long distance communication, therefore, increase of soliton wavelengths is recommended, where the sensitivity of the microring system can be obtained by generations of soliton pulses with larger FSR and smaller FWHM.

References

1. Nikoukar A, Amiri IS, Alavi SE, Shahidinejad A, Anwar T, Supa'at ASM, Idrus SM, Teng LY (2014) Theoretical and simulation analysis of the Add/Drop filter ring resonator based on the Z-transform method theory. The 2014 Third ICT International Student Project Conference (ICT-ISPC2014). Thailand, IEEE
2. Afroozeh A, Amiri IS, Zeinalinezhad A, Pourmand SE, Ahmad H (2015) Comparison of control light using Kramers-Kronig method by three waveguides. J Comput Theor Nanosci
3. Afroozeh A, Amiri IS, Chaudhary K, Ali J, Yupapin PP (2014) Analysis of optical ring resonator. Nova Science, Advances in Laser and Optics Research, New York
4. Shahidinejad A, Soltanmohammadi S, Amiri IS, Anwar T (2014) Solitonic pulse generation for inter-satellite optical wireless communication. Quantum Matter 3(2):150–154
5. Zeinalinezhad A, Pourmand SE, Amiri IS, Afroozeh A (2014) Stop light generation using nano ring resonators for ROM. J Comput Theor Nanosci
6. Afroozeh A, Amiri IS, Zeinalinezhad A (2014) Micro ring resonators and applications. LAP LAMBERT Academic Publishing, Saarbrücken, Germany
7. Afroozeh A, Amiri IS, Ali J, Yupapin PP (2012) Determination of Fwhm for solition trapping. Jurnal Teknologi 55:77–83
8. Afroozeh A, Amiri IS, Bahadoran M, Ali J, Yupapin PP (2012) Simulation of soliton amplification in micro ring resonator for optical communication. Jurnal Teknologi 55:271–277
9. Afroozeh A, Amiri IS, Jalil MA, Kouhnavard M, Ali J, Yupapin PP (2011) Multi soliton generation for enhance optical communication. Appl Mech Mater 83:136–140
10. Afroozeh A, Amiri IS, Kouhnavard M, Jalil M, Ali J, Yupapin PP (2010) Optical dark and bright soliton generation and amplification. AIP Conf Proc 1341:259–263
11. Afroozeh A, Bahadoran M, Amiri IS, Samavati AR, Ali J, Yupapin PP (2012) Fast light generation using GaAlAs/GaAs waveguide. Jurnal Teknologi 57:17–23
12. Ali J, Afroozeh A, Amiri IS, Hamdi M, Jalil M, Kouhnavard M, Yupapin PP (2010) Entangled photon generation and recovery via MRR. ICAMN, International Conference. Prince Hotel, Kuala Lumpur, Malaysia
13. Ali J, Afroozeh A, Amiri IS, Jalil M, Yupapin PP (2010) Wide and narrow signal generation using chaotic wave. Nanotech Malaysia, International Conference on Enabling Science & Technology. Kuala Lumpur, Malaysia
14. Ali J, Afroozeh A, Amiri IS, Jalil MA, Yupapin PP (2010) Dark and bright soliton trapping using NMRR. ICEM, Legend Hotel, Kuala Lumpur, Malaysia
15. Ali J, Amiri IS, Afroozeh A, Kouhnavard M, Jalil M, Yupapin PP (2010) Simultaneous dark and bright soliton trapping using nonlinear MRR and NRR. ICAMN, International Conference. Prince Hotel, Kuala Lumpur, Malaysia

16. Ali J, Amiri, IS, Jalil A, Kouhnavard A, Mitatha B, Yupapin PP (2010) Quantum internet via a quantum processor. International Conference on Photonics (ICP 2010). Langkawi, Malaysia
17. Ali J, Amiri IS, Jalil M, Kouhnavard M, Afroozeh A, Naim I, Yupapin PP (2010) Narrow UV pulse generation using MRR and NRR system. ICAMN, International Conference. Prince Hotel, Kuala Lumpur, Malaysia
18. Ali J, Jalil M, Amiri IS, Afroozeh A, Kouhnavard M, Naim I, Yupapin PP (2010) Multi-wavelength narrow pulse generation using MRR. ICAMN, International Conference. Prince Hotel, Kuala Lumpur, Malaysia
19. Ali J, Kouhnavard M, Afroozeh A, Amiri IS, Jalil MA, Yupapin PP (2010) Optical bistability in a FORR. ICEM, Legend Hotel, Kuala Lumpur, Malaysia
20. Ali J, Kouhnavard M, Amiri IS, Afroozeh A, Jalil MA, Naim I, Yupapin PP (2010) Localization of soliton pulse using nano-waveguide. ICAMN, International Conference. Prince Hotel, Kuala Lumpur, Malaysia
21. Ali J, Nur H, Lee S, Afroozeh A, Amiri IS, Jalil M, Mohamad A, Yupapin PP (2010). Short and millimeter optical soliton generation using dark and bright soliton. AMN-APLOC International Conference. Wuhan, China
22. Ali J, Raman K, Afroozeh A, Amiri IS, Jalil MA, Nawi IN, Yupapin PP (2010) Generation of DSA for security application. 2nd International Science, Social Science, Engineering Energy Conference (I-SEEC 2010). Nakhonphanom, Thailand
23. Ali J, Raman K, Kouhnavard M, Amiri IS, Jalil MA, Afroozeh A, Yupapin PP (2011) Dark soliton array for communication security. AMN-APLOC International Conference. Wuhan, China
24. Ali J, Roslan M, Jalil M, Amiri IS, Afroozeh A, Nawi I, Yupapin PP (2010) DWDM enhancement in micro and nano waveguide. AMN-APLOC International Conference. Wuhan, China
25. Shahidinejad A, Amiri, IS, Anwar, T (2014) Enhancement of Indoor wavelength division multiplexing-based optical wireless communication using microring resonator. Rev Theor Sci 2(3):201–210
26. Amiri IS, Afroozeh A, Ali J, Yupapin PP (2012) Generation of quantum codes using up and down link optical solition. Jurnal Teknologi 55:97–106
27. Amiri IS, Afroozeh A, Bahadoran M (2011) Simulation and analysis of multisoliton generation using a PANDA ring resonator system. Chin Phys Lett 28(10):104205
28. Amiri IS, Afroozeh A, Bahadoran M, Ali J, Yupapin, PP (2012) Molecular transporter system for qubits generation. Jurnal Teknologi 55:155–165
29. Amiri IS, Afroozeh A, Nawi IN, Jalil MA, Mohamad A, Ali J, et al. (2011) Dark soliton array for communication security. Procedia Eng 8:417–422
30. Amiri IS, Ahsan R, Shahidinejad A, Ali J, Yupapin PP (2012) Characterisation of bifurcation and chaos in silicon microring resonator. IET Commun 6(16):2671–2675
31. Amiri IS, Alavi SE, Idrus SM (2015) Introduction of fiber waveguide and soliton signals used to enhance the communication security. Soliton Coding for Secured Optical Communication Link. Springer, USA, pp 1–16
32. Amiri IS, Alavi SE, Idrus SM (2015) Results of digital soliton pulse generation and transmission using microring resonators. Soliton Coding for Secured Optical Communication Link. Springer, USA, pp 41–56
33. Amiri IS, Alavi SE, Idrus SM (2015) Theoretical background of microring resonator systems and soliton communication. Soliton Coding for Secured Optical Communication Link. Springer, USA, pp 17–39
34. Amiri IS, Ali J (2012) Generation of nano optical tweezers using an add/drop interferometer system. 2nd Postgraduate Student Conference (PGSC). Singapore
35. Amiri IS, Ali J (2013) Data signal processing via a manchester coding-decoding method using chaotic signals generated by a PANDA ring resonator. Chin Opt Lett 11(4):041901–041904
36. Amiri IS, Ali J (2013) Nano optical tweezers generation used for heat surgery of a human tissue cancer cells using add/drop interferometer system. Quantum Matter 2(6):489–493

37. Amiri IS, Ali J (2013) Optical buffer application used for tissue surgery using direct interaction of nano optical tweezers with nano cells. Quantum Matter 2(6):484–488
38. Amiri IS, Ali J (2013) Single and multi optical soliton light trapping and switching using microring resonator. Quantum Matter 2(2):116–121
39. Amiri IS, Ali J (2014) Characterization of optical bistability in a fiber optic ring resonator. Quantum Matter 3(1):47–51
40. Amiri IS, Ali J (2014) Picosecond soliton pulse generation using a PANDA system for solar cells fabrication. J Comput Theor Nanosci 11(3):693–701
41. Amiri IS, Ali J, Yupapin PP (2012) Enhancement of FSR and finesse using add/drop filter and PANDA ring resonator systems. Intern J Mod Phys B 26(04):1250034
42. Amiri IS, Khanmirzaei MH, Kouhnavard M, Yupapin PP, Ali J (2012) Quantum entanglement using multi dark soliton correlation for multivariable quantum router. In: Moran AM (ed) Quantum entanglement. Nova Science Publisher, New York, pp 111–122
43. Amiri IS, Nikoukar A, Ali J, Yupapin PP (2012) Ultra-short of pico and femtosecond soliton laser pulse using microring resonator for cancer cells treatment. Quantum Matter 1(2):159–165
44. Amiri IS, Nikoukar A, Shahidinejad A, Ali J, Yupapin PP (2012) Generation of discrete frequency and wavelength for secured computer networks system using integrated ring resonators. Computer and Communication Engineering (ICCCE) Conference, IEEE Explore, Malaysia
45. Amiri IS, Nikoukar A, Shahidinejad A, Ranjbar M, Ali J, Yupapin PP (2012) Generation of quantum photon information using extremely narrow optical tweezers for computer network communication. GSTF J Comput 2(1):140
46. Amiri IS, Raman K, Afroozeh A, Jalil MA, Nawi IN, Ali J, et al. (2011) Generation of DSA for security application. Procedia Eng 8:360–365
47. Amiri IS, Ranjbar M, Nikoukar A, Shahidinejad A, Ali J, Yupapin PP (2012) Multi optical soliton generated by PANDA ring resonator for secure network communication. Computer and Communication Engineering (ICCCE) Conference, IEEE Explore, Malaysia
48. Amiri IS, Shahidinejad A, Nikoukar A, Ranjbar M, Ali J, Yupapin PP (2012) Digital binary codes transmission via TDMA networks communication system using dark and bright optical soliton. GSTF J Comput 2(1):12
49. Amiri IS (2011) FWHM measurement of localized optical soliton. The International Conference for Nano materials Synthesis and Characterization, International Atomic Energy Agency (IAEA), Malaysia
50. Amiri IS (2011) Optical soliton trapping for quantum key generation. The International Conference for Nano materials Synthesis and Characterization, International Atomic Energy Agency (IAEA), Malaysia
51. Amiri IS (2014) Multiplex and de-multiplex of generated multi optical soliton by MRRs using fiber optics transmission link. Quantum Matter
52. Amiri IS, Afroozeh A (2014) Spatial and temporal soliton pulse generation by transmission of chaotic signals using fiber optic link advances in laser and optics research. Nova Science Publishers, New York
53. Amiri IS, Nikoukar A, Shahidinejad A, Anwar T, Ali J (2014) Quantum transmission of optical tweezers via fiber optic using half-panda system. Life Sci J 10(12s):391–400
54. Amiri IS, Nikoukar A, Shahidinejad A, Anwar T (2014) The proposal of high capacity GHz soliton carrier Signals applied for wireless commutation. Rev Theor Sci 2(4):320–333
55. Amiri IS, Nikoukar A, Ali J (2013) GHz frequency band soliton generation using integrated ring resonator for WiMAX optical communication. Opt Quantum Electron
56. Amiri IS, Shahidinejad A (2014) Generating of 57–61 GHz frequency band using a panda ring resonator. Quantum Matter
57. Amiri IS, Ali J (2013) Nano particle trapping by ultra-short tweezer and wells using MRR interferometer system for spectroscopy application. Nanosci Nanotechnol Lett 5(8):850–856
58. Amiri IS, Barati B, Sanati P, Hosseinnia A, Mansouri Khosravi HR, Pourmehdi S, et al. (2014) Optical stretcher of biological cells using sub-nanometer optical tweezers generated by an add/drop microring resonator system. Nanosci Nanotechnol Lett 6(2):111–117

59. Amiri IS, Ali J (2014) Femtosecond optical quantum memory generation using optical bright soliton. J Comput Theor Nanosci 11(6):1480–1485
60. Amiri IS, Ali J (2014) Generating highly dark-Bright Solitons by Gaussian Beam Propagation in a PANDA Ring Resonator. J Comput Theor Nanosci 11(4):1092–1099
61. Amiri IS, Ali J (2014) Optical quantum generation and transmission of 57–61 GHz frequency band using an optical fiber optics. J Comput Theor Nanosci 11(10)
62. Amiri IS, Ali J (2014) Simulation of the single ring resonator based on the Z-transform method theory. Quantum Matter 3(6):519–522
63. Amiri IS, Ebrahimi M, Yazdavar AH, Gorbani S, Alavi SE, Idrus SM, et al. (2014) Transmission of data with orthogonal frequency division multiplexing technique for communication networks using GHz frequency band soliton carrier. IET Commun 8(8):1364–1373
64. Amiri IS, Naraei P (2014) Optical transmission characteristics of an optical add-drop interferometer system. Quantum Matter
65. Amiri IS, Naraei P, Ali J (2014) Review and theory of optical soliton generation used to improve the security and high capacity of MRR and NRR passive systems. J Comput Theor Nanosci 11(9):1875–1886
66. Amiri IS, Alavi SE, Rahim FJ, Idrus SM (2014) Analytical treatment of the ring resonator passive systems and bandwidth characterization using directional coupling coefficients. J Comput Theor Nanosci
67. Amiri IS, Alavi SE, Idrus' SM (2014) Solitonic pulse generation and characterization by integrated ring resonators. 5th International Conference on Photonics. ICP2014. IEEE, Kuala Lumpur
68. Amiri IS, Alavi SE, Ali J (2013) High capacity soliton transmission for indoor and outdoor communications using integrated ring resonators. Int J Commun Sys
69. Amiri IS, Alavi SE, Bahadoran M, Afroozeh A, Idrus SM (2014) Nanometer bandwidth soliton generation and experimental transmission within nonlinear fiber optics using an add-drop filter system. J Comput Theor Nanosci
70. Amiri IS, Alavi SE, Idrus SM (2015) RF signal generation and wireless transmission using PANDA and Add/drop systems. J Comput Theor Nanosci
71. Amiri IS, Alavi SE, Idrus SM, Supa'at ASM, Ali J, Yupapin PP (2014) W-band OFDM transmission for radio-over-fiber link using solitonic millimeter wave generated by MRR. IEEE J Quantum Electron 50(8):622–628
72. Amiri IS, Alavi SE, Idrus SM, Nikoukar A, Ali J (2013) IEEE 802.15.3c WPAN standard using millimeter optical soliton pulse generated by a panda ring resonator. IEEE Photonics J 5(5):7901912
73. Amiri IS, Ghorbani S, Naraei P (2014) Chaotic carrier signal generation and quantum transmission along fiber optics communication using integrated ring resonators. Quantum Matter
74. Amiri IS, Soltanmohammadi S, Shahidinejad A, Ali J (2013) Optical quantum transmitter with finesse of 30 at 800-nm central wavelength using microring resonators. Opt Quantum Electron 45(10):1095–1105
75. Amiri IS, Nikoukar A, Alavi SE (2014) Soliton and radio over fiber (RoF) applications. LAP LAMBERT Academic Publishing, Saarbrücken, Germany
76. Amiri IS, Rahim FJ, Arif AS, Ghorbani S, Naraei P, Forsyth D, et al. (2014). Single soliton bandwidth generation and manipulation by microring resonator. Life Sci J 10(12s):904–910
77. Amiri IS, Alavi SE, Idrus SM, Afroozeh A, Ali J (2014) Soliton generation by ring resonator for optical communication application. Nova Science Publishers, New York
78. Amiri IS, Alavi SE, Idrus SM (2014) Soliton coding for secured optical communication link. Springer, USA
79. Amiri IS, Alavi SE, Idrus, SM, Kouhnavard, M (2014) Microring resonator for secured optical communication. Amazon, USA
80. Amiri IS, Alavi SE, Shahidinejad A, Nikoukar A, Anwar T, Supa'at ASM, Idrus SM, Yen NK (2014) Characterization of ultra-short soliton generation using MRRs. The 2014 Third ICT International Student Project Conference (ICT-ISPC2014). IEEE, Thailand

81. Kouhnavard M, Amiri IS, Jalil M, Afroozeh A, Ali J, Yupapin, PP (2010) QKD via a quantum wavelength router using spatial soliton. AIP Conf Proc 1347:210–216

82. Nikoukar A, Amiri IS, Ali J (2013) Generation of nanometer optical tweezers used for optical communication networks. Int J Innov Res Comput Commun Eng 1(1):77–85

83. Nikoukar A, Amiri IS, Shahidinejad A, Shojaei A, Ali J, Yupapin PP (2012) MRR quantum dense coding for optical wireless communication system using decimal convertor. Computer and Communication Engineering (ICCCE) Conference, IEEE Explore, Malaysia

84. Sanati P, Afroozeh A, Amiri IS, Ali J, Chua LS (2014) Femtosecond pulse generation using microring resonators for eye nano surgery. Nanosci Nanotechnol Lett 6(3):221–226

85. Ridha NJ, Mohamad FK, Amiri IS, Saktioto, Ali J, Yupapin PP (2010). Controlling center wavelength and free spectrum range by MRR radii. International Conference on Experimental Mechanics (ICEM). Kuala Lumpur, Malaysia

86. Alavi SE, Amiri IS, Idrus SM, Supa'at ASM (2014) Generation and wired/wireless transmission of IEEE802.16m signal using solitons generated by microring resonator. Opt Quantum Electron

87. Alavi SE, Amiri IS, Idrus SM, Supa'at ASM, Ali J, Yupapin PP (2014) All optical OFDM generation for IEEE802.11a based on soliton carriers using microring resonators. IEEE Photonics J 6(1)

88. Alavi SE, Amiri IS, Idrus SM, Supa'at ASM, Ali J (2013) Chaotic signal generation and trapping using an optical transmission link. Life Sci J 10(9s):186–192

89. Alavi SE, Amiri IS, Idrus SM, Supa'at ASM (2014) Optical amplification of tweezers and bright soliton using an interferometer ring resonator system. J Comput Theor Nanosci

90. Alavi SE, Amiri IS, Idrus SM, Ali, J (2013) Optical wired/Wireless communication using soliton optical tweezers. Life Sci J 10(12s):179–187

91. Sadegh Amiri I, Nikmaram M, Shahidinejad A, Ali J (2013) Generation of potential wells used for quantum codes transmission via a TDMA network communication system. Secur Commun Netw 6(11):1301–1309

92. Shahidinejad A, Nikoukar A, Amiri IS, Ranjbar M, Shojaei A, Ali J, Yupapin PP (2012) Network system engineering by controlling the chaotic signals using silicon micro ring resonator. Computer and Communication Engineering (ICCCE) Conference, IEEE Explore, Malaysia

93. Suwanpayak N, Songmuang S, Jalil MA, Amiri IS, Naim I, Ali J, et al. (2010) Tunable and storage potential wells using microring resonator system for bio-cell trapping and delivery. AIP Conf Proc 1341:289–291

94. Teeka C, Songmuang S, Jomtarak R, Yupapin PP, Jalil M, Amiri IS, et al. (2011) ASK-to-PSK generation based on nonlinear microring resonators coupled to one MZI arm. AIP Conf Proc 1341(1):221–223

Chapter 5
Conclusion

Abstract High capacity and secured optical communication signals are of importance in optical fiber communication. Ring resonators can be used for optical communication. In this book, the security and capacity performed by soliton signals for optical communications are investigated. The Optisystem and MATLAB softwares are used to simulate the results based on the Z-transform and iterative methods. Actual data from practical experiments are implemented for generating programming codes. The high capacity of the output signals is obtained by generation of multiple signals, available from channels and large bandwidth. Secured communication is performed using dark soliton pulse and chaotic signal with a spectrum of wavelengths. The chaotic signal, dark soliton, ultra-short pulses, soliton array are investigated for enhancing the security and capacity of optical communication systems. Therefore, the ring resonator system can be used to provide secured and high capacity for optical soliton communication signals using input Gaussian beam, bright and dark soliton pulses.

Keywords Secured optical communication · Optical fiber communication · Ring resonators · Ultra-short pulses · Soliton array · Bright and dark soliton pulses

A highly secured optical soliton signal has been developed for generating chaotic signals, dark soliton and ultra-short pulses. The ring resonator system has been able to provide a high capacity communication channel. The nonlinear effects of a single ring resonator such as optical bifurcation and chaos have been presented. Simulated results show that the bifurcation and chaos occur in different round-trip times and input power. Occurrence of bifurcation at lower input power or smaller round trip is a beneficial effect in order to improve the nonlinear microring system. System of series ring resonators connected to an add/drop filter system can be used to perform the dark and bright soliton generation. In order to generate multisoliton in the form of dark and bright soliton, a Gaussian beam is used as input pulse. The bright soliton is used for long distance communication, while the dark soliton is used to perform the security of the system. High capacity of

© The Author(s) 2015
I. Sadegh Amiri and A. Afroozeh, *Ring Resonator Systems to Perform Optical Communication Enhancement Using Soliton*, SpringerBriefs in Applied Sciences and Technology, DOI 10.1007/978-981-287-197-8_5

chaotic signals can be generated during propagation of light inside the three rings. The add/drop system filters the chaotic signals in which the multi spatial soliton of dark and bright pulses with FWHM and FSR of 31 GHz and 0.285 THz could be generated respectively. Bright soliton with power of 2 W is inserted into the ring system where generation of multi dark soliton pulses with FWHM of 50 pm and FSR of 1.44 nm could be obtained. The multisoliton pulses with FWHM and FSR of 10 and 490 pm are simulated respectively for the input bright soliton and Gaussian beam having power of 3 and 2 W.

Required signals of ultra-short soliton were perform which can be used in optical switching techniques. Single ultra-short temporal and spatial soliton pulses with FWHM of 4.8 ps, 17.5 pm, 712 fs, 0.34 nm, 83 fs and 84 fs could be generated, while the simulated multisoliton having FWHM and FSR of 16.53 ps, 19 pm, 140 ps, 40 pm and 502 ps, 576 pm, 3.6 ns and 1.45 nm respectively. The PANDA ring resonator is used to generate ultra-sharp bright and dark soliton pulses. These signals can be multiplexed through an optical multiplexer system and finally generate pulses of soliton arrays. Thus the bright soliton pulse with FWHM = 0.3 nm and dark soliton pulse with FWHM = 1.6 nm could be generated. Optical bright or dark soliton pulses could be generated and used to perform binary codes using PBS, where the entangle photon generation is performed within this device. Required binary codes could be generated after the ultra-short multiplexed signals with FWHM and FSR of 0.8 and 5 nm respectively were travelling into the PBS. The multisoliton pulse generation could be performed using a PANDA ring resonator system connected to an add/drop filter system. Generated chaotic signals from the PANDA system can be inserted into the add/drop filter system, thus multisoliton with FWHM and FSR of 425 pm, 50 pm and 1.145 nm, 1.44 nm are generated. In conclusion, the ring resonators system have been used and developed for increasing the channel capacity and secured communication.